Lecture Notes in Artificial

Subseries of Lecture Notes in Computer Science
Edited by J. G. Carbonell and J. Siekmann

Lecture Notes in Computer Science
Edited by G. Goos, J. Hartmanis and J. van Leeuwen

Michael Fisher   Richard Owens (Eds.)

Executable Modal and Temporal Logics

IJCAI '93 Workshop
Chambery, France, August 28, 1993
Proceedings

 Springer

Series Editors

Jaime G. Carbonell
School of Computer Science
Carnegie Mellon University
Pittsburgh, PA 15213-3891, USA

Jörg Siekmann
University of Saarland
German Research Center for Artificial Intelligence (DFKI)
Stuhlsatzenhausweg 3, D-66123 Saarbrücken, Germany

Volume Editors

Michael Fisher
Department of Computing, Manchester Metropolitan University
Chester Street, Manchester M1 5GD, United Kingdom

Richard Owens
Namura Research Institute Europe Ltd.
1 St Martin's-le-Grand, London EC1A 4NP, United Kingdom

CR Subject Classification (1991): I.2.3, D.1.6, F.4.1

ISBN 3-540-58976-7 Springer-Verlag Berlin Heidelberg New York

CIP data applied for

This work is subject to copyright. All rights are reserved, whether the whole or part of
the material is concerned, specifically the rights of translation, reprinting, re-use of
illustrations, recitation, broadcasting, reproduction on microfilms or in any other way,
and storage in data banks. Duplication of this publication or parts thereof is permitted
only under the provisions of the German Copyright Law of September 9, 1965, in its
current version, and permission for use must always be obtained from Springer-Verlag.
Violations are liable for prosecution under the German Copyright Law.

© Springer-Verlag Berlin Heidelberg 1995
Printed in Germany

Typesetting: Camera ready by author
SPIN: 10485359   45/3140-543210 - Printed on acid-free paper

Preface

The direct execution of logical statements, through languages such as PROLOG, has been both useful and influential within Computer Science and Artificial Intelligence. Yet, in recent years the requirement for greater expressive power has found languages based on first-order logic wanting. As a logical basis of knowledge representation, classical first-order logic has been superseded by, for example, *modal logics* for representing knowledge, belief, desire, etc, and *temporal logics* for representing temporal information and specifying and verifying reactive systems.

Just as the development of sophisticated theorem-proving techniques for first-order logic led to executable forms, such as PROLOG, so theorem-proving techniques for modal and temporal logics are being used in the development of executable forms of these logics. Each executable logic combines not only a *logical* perspective, but also an *operational* model, drawn from its intended application areas. Thus, though a variety of such languages have appeared, they exhibit a wide range of characteristics and execution mechanisms.

This volume contains updated and extended versions of papers presented at the Workshop on 'Executable Modal and Temporal Logics' held as part of IJCAI'93. These papers describe a range of approaches, not only from a logical point of view, but also from programming language and applications standpoints. As such, we believe this volume provides an indication of the breadth of research activity within this expanding and exciting field.

Table of Contents

An Introduction to Executable Modal and Temporal Logics

Michael Fisher & Richard Owens

Abstract. In recent years a number of programming languages based upon the direct execution of either modal or temporal logic formulae have been developed. This use of non-classical logics provides a powerful basis for the representation and implementation of a range of dynamic behaviours. Though many of these languages are still experimental, they are beginning to be applied, not only in Computer Science and AI, but also in less obvious areas such as process control and social modelling.

This paper provides an introduction to some of the basic concepts of executable modal and temporal logics.

1 Introduction

In this paper we provide an introduction to the basic concepts involved in the direct execution of modal and temporal logics. Although previously seen as esoteric, this area of research is beginning to expand, with a variety of languages being developed and applied. We do not expect the reader to gain a deep understanding of the mechanisms for executing modal and temporal logics from this paper; our intention is simply to introduce the underlying concepts, and to provide a set of references that can be followed if desired. Consequently, along with an outline of the logics and some of the common execution methods used, we provide a brief survey of implemented languages based upon the execution of modal or temporal logics. For the sake of simplicity, we restrict our description to the propositional versions of the logics and their execution mechanisms.

1.1 Motivation

First, we consider the question of why we might want to directly execute logics *at all*.

Why Execute Logical Formulae? In general, what does it mean to execute a formula, F, of logic, L? In logical terms, the execution of a formula corresponds to building a model for that formula. Thus, the model constructed, call it $\mathcal{M}$, satisfies the relation

$$\mathcal{M} \models_L F$$

In the simplest case, execution can be carried out using basic theorem-proving techniques. However, if external constraints are placed upon F, then $\mathcal{M}$ must be constructed in a slightly different way (e.g., by some interaction with an environment). Note that languages such as Prolog effectively build a model for the formula by attempting to prove the negation of a goal.

Also, as F represents a declarative statement, then producing $\mathcal{M}$ can be seen as execution within the declarative language L. Further, if F is a specification, then constructing $\mathcal{M}$ can also be seen as prototyping an implementation of that specification.

Although the logic might be expensive to reason in, it has been shown, for example in Prolog, that efficiency can be gained by specialising the theorem-proving mechanism. Although this may introduce some non-logical operational constraints into the execution mechanism, the language still retains its desirable properties, such as its declarative, high-level descriptive capabilities.

Why Execute Modal Logic? Many problems, particularly in knowledge representation, involve the division of the problem space into a number of sub-spaces. For example, in representing problems involving distributed reasoning, the total knowledge of the system might be the sum of the knowledge of its components. Alternatively knowledge might only be propagated through the relationships between the sub-spaces — for example local groups of components may share knowledge which is not made available to remote components.

Classical first-order logic expresses relationships between terms representing members of a domain. The truth values of statements in the logic are derived with respect to a particular interpretation of the symbols of the logic to members of some domain. Because there is a single domain, and one interpretation into that domain, classical logic is restricted to statements about a *global* problem space.

This is a strength because, since the problem space can be arbitrarily complex, classical first-order logic can express any first-order condition over arbitrarily complex problem spaces. However, it is also a weakness, as there are no language constructs for handling structural complexity within the problem space.

While the universal nature of classical first-order logic allows such problems to be specified, statements about the general underlying problem space must be incorporated into statements relating to the particular problem being considered. For example, we might state that "for all members m of the domain, if m is an agent, and m is connected to a particular agent a, then if a knows that P is true, m knows that P is true". It is far more preferable if we state that (in the context of agent a) "if P is known to be true then all connected agents know that P is true".

Any theorem-proving or execution in such structured problem spaces would ideally have knowledge of the structure, and exploit the structure to optimise the execution.

Without a means to make such structural optimisations, much effort is spent in repeatedly proving axioms about the structure, which are already known.

Modal logics offer a means to overcome such problems by use of additional connectives available in the logic (the *modalities*) to represent aspects of the structure of the problem space, and by providing theorem-proving techniques which utilise properties of the structure of the problem space to optimise deductions.

Why Execute Temporal Logic? Given that executing logical statements (both classical and modal) has some benefits, then why do we choose to execute *temporal* logic?

Perhaps the main reason is to provide a high-level language with significant expressive power. Temporal logics are useful for reasoning about a changing world and, as the temporal order of actions/events can be described, they are useful for representing dynamic systems. Consequently, temporal logics are widely used in the specification and verification of *reactive* systems [53], and in applications where the concept of time is central, such as temporal planning [4], temporal representation [6] and temporal databases [22]. Thus, programming languages that provide access to such temporal concepts have a wide range of applications.

Further, the logical power of the language allows us to express complex temporal properties of systems. For example, if *discrete* temporal logics are used, where time is represented as a sequence of distinct moments, then temporal formulae can effectively represent the individual steps of an execution. Similarly, by using temporal logics, we can describe high level properties of systems, e.g., some condition occurring *sometime* in the future, or occurring *infinitely often* (c.f., liveness and fairness).

Finally, as temporal logics are complex to reason with [62], then any formal specification and refinement technique is likely to encounter problems in verifying program transformations. If we are able to execute the temporal specifications directly, then this avoids the necessity of proving a large number of refinements correct (although the cost of execution itself is likely to be high).

1.2 Structure of these notes

We begin our discussion by briefly outlining, in §2, the basic principles of modal logics. In §3, we outline the predominant mechanism for executing modal logics, based upon logic programming using modal resolution. As a wider range of executable temporal logics have been developed, we consider two alternative temporal logics, namely *discrete* and *interval* temporal logics in §4, both of which provide bases for execution. A range of execution options for these temporal logics are described in §5. Finally, in §6, we provide some concluding remarks.

2 Modal Logics

Modern modal logics extend classical logics with (generally) a new connective $\Box$ and its derivable counterpart $\Diamond$. These are known as *necessity* and *possibility* respectively. If a formula $\Box p$ is true, it means that p is necessarily true, i.e. true in every possible scenario, and $\Diamond p$ means that p is possibly true, i.e. true in at least one possible scenario. We can define $\Diamond$ in terms of $\Box$:

$$\Diamond p \Leftrightarrow \neg \Box \neg p$$

so that p is possible exactly when its negation is not necessarily true.

In order to give meaning to $\Box$ and $\Diamond$, models for modal logics are usually based on 'possible worlds', which are essentially a collection of connected models for classical logic. The possible worlds are linked by a relation which determines which worlds are accessible from any given world. It is this *accessibility relation* which determines the nature of the modal logic.

Each world is given a unique label, taken from a set S, which is usually countably infinite. The accessibility relation R is a binary relation on S. The pairing of S and R defines a *frame* or structure which underpins the models for the modal logic. To complete the model we add an interpretation, $h: S \times PROP \rightarrow \{\text{true, false}\}$, of symbols in each state so that

$$\langle S, R, h \rangle \models_s a \quad \text{iff} \quad h(s, a) = \text{true}.$$

This is read as a is true in the model $\langle S, R, h \rangle$ in world s iff h maps a to true in world s. In general when a formula φ is true in a world s in a model $\mathcal{M}$ it is denoted by

$$\mathcal{M} \models_s \varphi$$

and if it is true in every world in the set S, it is said to be true in the model, and denoted

$$\mathcal{M} \models \varphi$$

The boolean connectives are given the usual meaning:

$$\langle S, R, h \rangle \models_s \varphi \vee \psi \quad \text{iff} \quad \langle S, R, h \rangle \models_s \varphi \text{ or } \langle S, R, h \rangle \models_s \psi.$$
$$\langle S, R, h \rangle \models_s \varphi \Rightarrow \psi \quad \text{iff} \quad \langle S, R, h \rangle \models_s \varphi \text{ implies } \langle S, R, h \rangle \models_s \psi.$$

The frame enters the semantic definition only when the modality $\Box$ is used, as the formula $\Box \varphi$ is true in a world s exactly when every world t in S which is accessible from s (i.e. we have sRt) has φ true. More formally,

$$\langle S, R, h \rangle \models_s \Box \varphi \quad \text{iff} \quad \text{for all } t \in S, \quad sRt \text{ implies } \langle S, R, h \rangle \models_t \varphi$$

Examples of modal formulae To illustrate the use of modalities to represent semantic structure, we give a few examples:

- If, in all alternative scenarios, I am rich then I am currently rich: $\Box rich \Rightarrow rich$
- Either p has to be true or p has to be false: $\Box p \vee \Box \neg p$ (note that this is distinct from just $p \vee \neg p$, because we are insisting that either p is true in every possible scenario, or it is false in every possible scenario).
- If I am rich then it is necessarily possible for me to be rich: $rich \Rightarrow \Box \Diamond rich$.

For further references to modal logics, and correspondences between modal axioms and the structure of frames, see [48, 18, 67].

3 Programming with Modal Logics

As discussed above, we wish to execute modal logic in preference to classical logic in some situations, because it is easier and more natural to describe systems which involve notions such as distributed reasoning, knowledge, and need in a language which has additional syntactic units (the modalities) to abstract away generic problem space structure. Execution mechanisms can be tailored to exploit specific meanings given to those modalities.

Execution of modal logics has progressed along a similar path as execution of classical logic, namely by developing mechanical proof procedures, and turning them into automated deduction techniques. This has often involved extending existing classical proof procedures to handle modalities, for example semantic tableaux have been extended to contain representations of possible worlds [39], where the extension is based on the use of the tableaux expansion rules for universal and existential quantification. Connection matrices, derived from tableaux have also been extended to modal logics [68].

Below we consider an approach to execution based upon the logic programming paradigm.

3.1 Modal Logic Programming — MOLOG

Not surprisingly, the predominant approach to the execution of modal logics is based upon standard logic programming. This is extended so that *modal* Horn Clauses are defined, based upon the clauses used in modal resolution, while the goal reduction process itself is based upon modal, rather than classical, resolution.

In order to provide a little background to this approach, we will outline the modal resolution method upon which modal logic programming is typically based.

Resolution and Modal Logics Resolution theory for classical logic has been extended to modal logics [25]. The conjunctive normal form on which resolution is based is expanded in modal resolution to be conjunctions of disjunctions of the form

$$L_1 \vee \cdots \vee L_{n_1} \vee \Box D_1 \vee \cdots \vee \Box D_{n_2} \vee \Diamond A_1 \vee \cdots \vee \Diamond A_{n_3}$$

where the L_i are literals (atoms or negated atoms), the D_i are formulae in the form of the disjunction above. The A_i are formulae in conjunctive normal form. All variables are universally quantified.

Additional resolution rules are required to handle the modalities. In particular, due to the complexity of modal logics, modal resolution resolvents may be obtained from a single clause. This is known as *unary* resolution. For example, the following formulae, both in the normal form, are logically equivalent:

$$A \vee \Diamond((A \vee \Box B) \wedge \neg A \wedge \Diamond C)$$
$$A \vee \Diamond(\Box B \wedge \Diamond C)$$

The A and $\neg A$ in the scope of the $\Diamond$ can be used as the basis of the resolution, to obtain the second formula as a resolvent of the first.

The modal extension of the classical resolution rule takes two forms. The first resolves on two $\square$ modalities: if we have two normal formula of the form $\square E$ and $\square F$ we will have E and F together in each accessible world. Therefore if there is a resolvent obtainable from E and F, that resolvent will be in every possible world. For example,

$$\square(A \vee B)$$
$$\square(\neg A \vee C)$$

will resolve on A and $\neg A$ to produce $\square(B \vee C)$ as resolvent. The second modal extension of the classical resolution rule resolves on a $\square$ and a $\Diamond$ modality: If we have two normal formula of the form $\square E$ and $\Diamond F$ we will have E in every accessible world, and F in at least one accessible world. Therefore if there is a resolvent obtainable from E and F that resolvent will be in at least one possible world. For example,

$$\square(A \vee B)$$
$$\Diamond(\neg A \vee \neg D)$$

will resolve on A and $\neg A$ to produce $\Diamond(B \vee \neg D)$ as resolvent.

We can summarise the unary resolution (denoted Γ), and the modal form of classical resolution (denoted Σ) by the definition below:

1. $\Sigma(A, \neg A) = \{\}$ (empty clause)
2. $\Sigma(D_1 \vee D_2, F) = \Sigma(D_1, F) \vee D_2$
3. $\Sigma(D_1 \wedge F_1, D_2 \wedge F_2) = \Sigma(D_1, D_2) \wedge F_1 \wedge F_2$
4. $\Sigma(\square E, \square F) = \square \Sigma(E, F)$
5. $\Sigma(\square E, \Diamond F) = \Diamond(E, F)$
6. $\Gamma(\Diamond(D_1 \wedge D_2 \wedge F)) = \Diamond(\Sigma(D_1, D_2) \wedge F)$

Different modal logics are obtained by varying the definitions of Σ and Γ.

A non-clausal form of modal resolution was developed by Abadi and Manna [2]. It is based on three sets of rules. A resolution rule yields resolvents by creating a disjunction of two formulae which have sub-formulae in common, and replacing the common sub-formula by 'true' in one disjunct, and 'false' in the other. Simplification rules are used to remove falsities from formulae, and to push negations onto atomic formulae. Finally deduction rules introduce logical deductions into the set of formulae being resolved. It is the members of this latter set that determine the precise nature of the modal logic. Deductions in the non-clausal modal resolution are non-trivial, and the method used to handle variables and quantification in the first-order case requires paramodulation to handle equalities that arise.

We now turn to a particular programming language based upon this approach. While, in general, few executable modal logics have been developed (though see [23, 20]), a large body of work has been carried out at IRIT developing MOLOG, a modal logic programming language of the the above form [26], and mechanisms for the extension of standard logic programming to non-classical logics [9, 8].

MOLOG itself uses modal Horn Clauses derived from the above clausal form, together with goal reduction rules based upon Modal SLD-resolution. For example,

$$\frac{\begin{array}{c} \leftarrow \Diamond p, G \\ \Box p \leftarrow B \end{array}}{\leftarrow B, G}$$

In extending this work, the group at Toulouse have developed a framework for the generation of inference mechanisms based upon non-classical extensions to logic programming, called the *Toulouse Inference Machine* [9], together with an abstract machine model for the implementation of these mechanisms, called TARSKI [8].

4 Temporal Logics

We now turn to temporal logics. As the languages described later are generally based on two different forms of temporal logic, namely *discrete* temporal logics and *interval* temporal logics, we (briefly) describe each of these separately.

4.1 Discrete Temporal Logics

This section describes a standard propositional temporal logic [24], called PTL, based on a discrete, linear model of time, with a finite past and infinite future [42]. Here, time can be visualised as a sequence of discrete moments, starting with a distinguished 'beginning of time' and extending infinitely into the future.

The language of PTL can be seen as that of classical logic extended with various modalities. Although this obviously has close links with modal logic, it can be seen as a particular form of modal logic whereby the reachability relation between worlds is taken as an earlier/later (i.e. temporal) relation. In addition to the '$\Diamond$' and '$\Box$' operators used in modal logics, PTL also uses the *next-time* operator, '$\bigcirc$'. In PTL, the intuitive meaning of these connectives is as follows:

- $\Diamond A$ is satisfied now if A is satisfied *sometime* in the future;
- $\Box A$ is satisfied now if A is satisfied *always* in the future;
- $\bigcirc A$ is satisfied now if A is satisfied at the *next* moment in time.

We can also introduce other operators, such as the *until* operator, '$\mathcal{U}$', and its weakened version, the *unless* operator, '$\mathcal{W}$'. The intuitive meaning of these binary connectives is:

- $A\mathcal{U}B$ is satisfied now if B is satisfied at some future moment, and A is satisfied until then;
- $\mathcal{W}$ is a binary connective similar to $\mathcal{U}$, allowing for the possibility that the second argument never becomes satisfied.

Intuitively, the models for PTL formulae are based on discrete, linear structures having a finite past and infinite future, i.e. sequences of the form

$$s_0, \ s_1, \ s_2, \ s_3, \ \dots$$

where each s_i, called a *state*, provides a propositional valuation. However, rather than representing the model structure in this way, we will define a model, σ, as

$$\sigma = \langle N, \pi_p \rangle$$

where

   N is the Natural Numbers, which is used to represent the sequence $s_0, s_1, s_2, s_3, \ldots$, and,

   π_p is a map from $N \times \mathcal{L}_p$ (the set of all proposition symbols) to $\{T, F\}$, giving a propositional valuation for each state in the sequence.

An interpretation for this logic is defined as a pair $\langle \sigma, i \rangle$, where σ is the model and i the index of the state at which the temporal statement is to be interpreted.

   A semantics for well-formed temporal formulae is a relation between interpretations and formulae, and is defined inductively as follows, with the (infix) semantic relation being represented by '$\models$'. The semantics of a proposition is defined by the valuation given to it at a particular state:

$$\langle \sigma, i \rangle \models p \quad \text{iff} \quad \pi_p(i, p) = T \qquad [\text{for } p \in \mathcal{L}_p].$$

The semantics of the standard propositional connectives is as in classical logic, e.g.,

$$\langle \sigma, i \rangle \models A \vee B \quad \text{iff} \quad \langle \sigma, i \rangle \models A \quad \text{or} \quad \langle \sigma, i \rangle \models B.$$

The semantics of the unary future-time temporal operators is defined as follows.

$$\langle \sigma, i \rangle \models \bigcirc A \quad \text{iff} \quad \langle \sigma, i+1 \rangle \models A$$
$$\langle \sigma, i \rangle \models \Diamond A \quad \text{iff} \quad \text{there exists } j \in N \text{ such that } j \geq i \text{ and } \langle \sigma, j \rangle \models A$$
$$\langle \sigma, i \rangle \models \Box A \quad \text{iff} \quad \text{for all } j \in N, \text{ if } j \geq i \text{ then } \langle \sigma, j \rangle \models A$$

Additionally, the syntax includes two binary future-time temporal operators, $\mathcal{U}$ and $\mathcal{W}$. The first of these is interpreted as follows, the second is an obvious weakening of this.

$$\langle \sigma, i \rangle \models A \mathcal{U} B \quad \text{iff} \quad \text{there exists } k \in N, \text{ such that } k \geq i \text{ and } \langle \sigma, k \rangle \models B \text{ and}$$
$$\text{for all } j \in N, \text{ if } i \leq j < k \text{ then } \langle \sigma, j \rangle \models A$$

$$\langle \sigma, i \rangle \models A \mathcal{W} B \quad \text{iff} \quad \langle \sigma, i \rangle \models A \mathcal{U} B \quad \text{or} \quad \langle \sigma, i \rangle \models \Box A$$

Finally, we also introduce the '**start**' operator, which is used to distinguish the unique "beginning of time", as follows.

$$\langle \sigma, i \rangle \models \textbf{start} \quad \text{iff} \quad i = 0$$

Examples of PTL Formulae To give some idea of the type of properties that can be represented in PTL, we provide the following (simple) examples.

- A simple choice of what should occur next: $p \wedge (\bigcirc q \vee \bigcirc r)$
- Using the *until* operator to represent intervals of time: *nervous* $\mathcal{U}$ *started*
- Using the *unless* operator: *poor* $\mathcal{W}$ *lottery-win*

- Temporal indeterminacy through the *sometime* operator: *born* $\Rightarrow \Diamond(rich \wedge happy)$
- Links with the representation of standard (imperative) computation:

$$x = 0 \wedge \bigcirc(x = x + 1) \wedge \ldots$$

- Something occurring infinitely often: $\Box \Diamond tick$
- One form of fairness: $\Box \Diamond ask \Rightarrow \Box \Diamond receive$
- Another (weaker) form of fairness: $\Box \Diamond ask \Rightarrow \Diamond receive$
- Induction:

$$[rich \Rightarrow happy \wedge \Box(happy \Rightarrow \bigcirc happy)] \Rightarrow [rich \Rightarrow \Box happy]$$

For further references to temporal logics, seen both from Logic and Computer Science viewpoints, see [19, 52, 24, 53, 48, 18, 67].

4.2 Interval Temporal Logic

While PTL is essentially based upon the notion of points, the interval temporal logic introduced here is concerned with the truth of statements over *intervals*. This logic is called ITL and was originally developed by Moszkowski in order to model digital circuits [56]. Its syntax contains the basic temporal operators of PTL, together with the *chop* operator, ';', which is used to concatenate intervals.

The semantics of ITL are given over a sequence, σ, as defined for PTL. However, statements are interpreted in a sub-sequence of, rather than at a point within, σ. Thus, the semantics of ITL operators can be given as below.

$\langle \sigma_b, \ldots, \sigma_e \rangle \models P \wedge Q$ iff $\langle \sigma_b, \ldots, \sigma_e \rangle \models P$ and $\langle \sigma_b, \ldots, \sigma_e \rangle \models Q$

$\langle \sigma_b, \ldots, \sigma_e \rangle \models P \vee Q$ iff $\langle \sigma_b, \ldots, \sigma_e \rangle \models P$ or $\langle \sigma_b, \ldots, \sigma_e \rangle \models Q$

$\langle \sigma_b, \ldots, \sigma_e \rangle \models \Box P$ iff for all $b \leq i \leq e$. $\langle \sigma_i, \ldots, \sigma_e \rangle \models P$

$\langle \sigma_b, \ldots, \sigma_e \rangle \models \bigcirc P$ iff $e > b$ and $\langle \sigma_{e+1}, \ldots, \sigma_e \rangle \models P$

$\langle \sigma_b, \ldots, \sigma_e \rangle \models P; Q$ iff exists $b \leq i \leq e$. $\langle \sigma_b, \ldots, \sigma_i \rangle \models P$ and $\langle \sigma_i, \ldots, \sigma_e \rangle \models Q$

In addition, '$\Diamond$' can be derived, this time in terms of ';', i.e.

$$\Diamond P \Leftrightarrow \textbf{true}; P$$

meaning that there is some sub-interval in which **true** is satisfied that is followed by a sub-interval in which P is satisfied.

Although full ITL is undecidable [56], both the languages described below use a subset of ITL called *Local ITL*, where the truth of propositions does not depend upon their values at the end of an interval [51].

For further references to interval temporal logics and their applications in such diverse areas and AI and Hardware Verification, see [5, 64, 58, 7]

5   Programming with Temporal Logics

While there have been relatively few executable modal logics produced, many research-
ers have developed executable temporal logics. Unfortunately, there are two major
problems associated with the execution of (discrete) temporal logics — complexity
and incompleteness.

1. Executing PTL is complex (PSPACE-complete [62]), while executing first-order
   temporal logic is undecidable (and *highly* complex [54]). Consequently, two approaches to the execution of temporal logic have been followed: to restrict the logic
   and provide a more efficient execution mechanism for the restricted fragment; or to
   execute the full logic and to embed more efficient heuristics within the execution
   mechanism.
2. First-order (discrete) temporal logic is incomplete [65, 3], so not every formula can
   be successfully executed. Again the same two approaches have been followed: to
   restrict the logic; or to disregard completeness and to view execution as simply an
   attempt to build a model for the formula.

Thus, as Merz [54] suggests, "the design of a temporal logic based programming language involves a tradeoff between expressiveness and (efficient) implementability".

The particular approaches that we outline below exemplify two alternative approaches. In §5.1, a restriction of PTL that follows the standard logic programming
paradigm is described. Logic programming has been successfully applied in a wide
variety of areas. Also, not only do standard logic programming languages, such as Prolog, satisfy desirable formal properties, but it has been shown that relatively efficient
implementations can be produced for these languages. Although logic programming
does not utilise the full power of classical logics, it comprises a simple fragment of
the logic (Horn Clauses), together with operational rules based upon standard resolution. Thus, once resolution rules for temporal logics were developed [2], it was not
surprising that temporal logic programming languages were developed. Unfortunately,
not only is the fragment of temporal logic that is used very restrictive (even more so
than Horn Clauses), but efficient implementations are difficult to develop.

An alternative approach, outlined in §5.2, is to ignore the problem of completeness and implement the full logic directly. In doing this it was recognised that certain
temporal formulae would be expensive to execute. However, many common temporal
'idioms' fit into this style of execution and can be implemented relatively efficiently.

While Interval Temporal Logics, such as ITL do not exhibit the problems of PTL
(or at least not to the same degree), two similar approaches to the execution of ITL
have been followed. These are briefly outlined in §5.3 and §5.4.

5.1   Temporal Logic Programming — TEMPLOG

As with modal logics, discrete temporal logics can be executed using the logic programming paradigm. Unfortunately, due to incompleteness of first-order temporal logic, the Horn Clause fragment of temporal logic is still too powerful to satisfy the
desirable properties of logic programming, so such a fragment must be restricted even

further [1, 14, 3]. Hence the search for a subset of temporal Horn Clauses logic that can be executed successfully using the logic programming paradigm.

The predominant approach to temporal logic programming is TEMPLOG [1]. Here, Temporal Horn Clauses can be categorised as either *initial* clauses (effectively if they contain **start**), or global clauses, and are restricted still further using the following constraints.

1. The '$\Diamond$' operator can only occur in the either the bodies of rules or in goals.
2. The '$\Box$' operator can only occur in what are termed, *initial definite permanent* clauses, which can be characterised as

$$\Box e \leftarrow d, \mathbf{start}, c.$$

These restrictions basically ensure that no $\Diamond$-formulae appear in the program clauses. Thus, the idea behind these restrictions is to allow goals and bodies of program rules to consist of formulae such as p, $\bigcirc q$ and $\Diamond r$, while heads of program rules may consist of formulae such as p, $\bigcirc q$ and $\Box r$.

Given a particular goal, together with a set of rules, the goal can be successively reduced using reduction rules based upon temporal resolution [2]. These are the temporal analogue of the standard resolution rule and are termed *Temporal SLD* (TSLD) resolution rules [2, 1]. In a simplified form, some of these rules are:

$$\frac{\leftarrow \bigcirc p, G \qquad p \leftarrow B}{\leftarrow \bigcirc B, G} \qquad \frac{\leftarrow \Diamond p, G \qquad p \leftarrow B}{\leftarrow \Diamond B, G} \qquad \frac{\leftarrow \Diamond p, G \qquad \bigcirc p \leftarrow B}{\leftarrow \Diamond B, G}$$

Properties of TEMPLOG There are various technical results relating to TEMPLOG, which can be summarised as follows.

Expressive Power Baudinet [14, 13] provides two alternative formulations of (first-order) TEMPLOG's declarative semantics and, by relating these to TEMPLOG's operational semantics (as given by the TSLD rules above), shows that these rules are complete. Further, she shows how propositional TEMPLOG programs represent a fragment of the temporal fixpoint calculus [10]. These important results show that, not only does TEMPLOG represent a (relatively) efficient execution mechanism for a fragment of PTL, but that, as in standard logic programming where a variety of semantic definitions coincide, both minimal model and fixpoint semantics coincide with TEMPLOG's operational semantics.

Characterising TEMPLOG The basic TEMPLOG form of temporal Horn Clause, together with its operational model based upon TSLD-resolution, can be characterised in a variety of different forms:

1. Orgun and Wadge [60, 59], describe a fragment of temporal Horn Clauses (called CHRONOLOG) that is expressively equivalent to TEMPLOG and that can be seen as a form of intensional logic programming. This fragment includes all the restrictions

of TEMPLOG but, in addition, restricts clauses so that no '◊' operators are allowed in the body of any clause and no ' □' operators are allowed in *any* heads. CHRONOLOG is equivalent to the restricted from of TEMPLOG called TL1 [1, 14].

2. Brzoska [16] describes the relationship between TEMPLOG and a form of constraint logic programming. In particular, he shows that TEMPLOG programs can be considered as *CLP(A)* programs over a suitable algebra *A*. This involves defining the particular algebra *A*, providing a translation from temporal Horn Clauses, and TEMPLOG clauses in particular, to elements of *A*, and showing that TSLD-resolution is a (restricted) form of the CLP-derivation mechanism over *A*.

3. Balbiani *et. al.* [9] describe the Toulouse Inference Machine (TIM), which provides a meta-language for defining extensions of logic programming based upon non-classical logics. The basic idea is to add the notion of *contexts* to standard logic programming and allow meta-rules to manipulate these. In particular, they show how the TSLD-resolution rules can be coded in the meta-language and hence how TEMPLOG can be implemented on top of Prolog.

5.2 Imperative Future — METATEM

The idea behind temporal logic programming, and TEMPLOG in particular, is to identify a subset of temporal logic that can not only be efficiently executed, but provides properties analogous to those found in standard logic programming. An alternative view is to attempt to execute *any* formula that can be written in temporal logic. This means that certain temporal constructs (e.g., '◯') are efficiently implementable, while others (e.g., '◊') may involve complex computation.

In order to do this, we abandon the logic programming paradigm as it restricts the form in which logical formulae can be represented and constrains their execution. Thus, rather than using an execution mechanism based upon resolution and refutation, as in logic programming, we now look at one based upon model construction techniques related to analytic tableaux. In particular, the general idea behind the *imperative future* approach [43] is to rewrite each formula to be executed into a set of formulae of the form

$$\text{condition on the past} \Rightarrow \text{constraint on the future}$$

then treat such formulae as templates showing how the future can be constructed given the past constructed so far. Thus, the term *imperative future* derives from the general idea that in order to construct the next state in the model, we check conditions on the previous states and, if necessary, apply these constraints to the construction of the next and future states.

Initially, Gabbay developed the language USF [43], which provides a small set of temporal operators ('until', 'since', and 'sometime') and relies upon the Separation Theorem [50, 43] to derive executable formulae of the above form. The METATEM language [11], is a development of USF consisting of a larger range of operators, a better defined execution mechanism [31] and a more practical normal form [30] derived from the normal form used in clausal temporal resolution [36, 37].

The basic idea behind METATEM [11, 31] is to directly use the formula to be executed in order to build a model (in the case of PTL, a sequence) for the formula. The formula to be executed is transformed into a normal form consisting of *rules* of the form

$$\textbf{start} \Rightarrow \bigvee_j l_j$$

or

$$\bigwedge_k l_k \Rightarrow \bigcirc \left(\bigvee_l t_l \vee \bigvee_n \Diamond t_n \right)$$

where l_j, l_k, t_l, and t_n are literals.

The intuition behind this form of rules is that the former (called *initial* rules) provide constraints upon the initial state, while the latter (called *global*) rules provide constraints upon the *next* state.

Examples of METATEM Rules Below are some simple examples showing some of the properties that might be represented directly as METATEM rules.

- Specifying initial conditions: **start** $\Rightarrow$ *sad*
- Specifying initial goals (eventualities):   **start** $\Rightarrow$ $\Diamond happy$
- Introducing permanent properties: *rich* $\Rightarrow$ $\Box happy$
- Defining transitions between states: (*sad* $\wedge$ $\neg rich$) $\Rightarrow$ $\bigcirc sad$
- Introducing new goals: ($\neg resigned$ $\wedge$ *sad*) $\Rightarrow$ $\Diamond famous$

METATEM Execution The execution of a METATEM program is an iterative process of labelling a model structure with the propositions true in each state, which eventually yields a model for the formula (if the formula is satisfiable). The model structure produced is a sequence of states, with an identified start point. The initial model may be empty, or it may contain some information already. In either case, execution starts at the initial state 0, and steps through each state in the structure in turn.

Thus, if we wish to construct the initial state, the inital rules are consulted, while if we wish to construct any other state, we generate constraints on this state from the global rules whose antecedents are satisfied in the previous state. As the $\Diamond$ operator is non-deterministic, the execution mechanism has a choice as to which states in the model to label in order to make the formula true. We can express this choice by the equivalence:

$$\Diamond a \Leftrightarrow (a \vee \bigcirc \Diamond a)$$

So to make $\Diamond a$ true, the interpreter can either

1. make a true immediately, or,
2. make $\Diamond a$ true in the next state.

The latter is achieved by passing a commitment $\Diamond a$ from the current state to the next state to be conjoined with the consequents of the successful rules.

Thus, the interpreter has a strategy not only for choosing between disjuncts, but also for choosing when to satisfy formulae of the form $\Diamond a$ (called *eventualities*). In

METATEM, the execution mechanism attempts to satisfy as many eventualities as possible. In the case of conflicting eventualities, e.g., $\Diamond a$ and $\Diamond \neg a$, the oldest outstanding eventuality is satisfied first. Any unsatisfied eventualities are passed on to the next state.

If a contradiction is generated within a state, i.e., execution of the rules has forced us to make both a proposition and its negation true in the current state, then a form of 'backtracking' — undoing previous choices — occurs. This backtracking undoes the model construction and and returns the execution to a previous choice point. If there are no more choice points left, the execution fails, signifying that the program is unsatisfiable.

Applications Although much of the development of METATEM has been suspended in favour of Concurrent METATEM (see below), the language has applications in system modelling [27], databases [28] and meta-level representation and planning [12].

5.3 Imperative Interval Temporal Logic — Tempura

Tempura [57, 45] is an executable temporal logic based upon the forward chaining execution of ITL. In this sense it was the precursor of the METATEM family of languages and provides a simpler and more tractable alternative to these approaches. In addition to the ITL operators outlined above, Tempura incorporates various derived constructs, often with connotations in imperative programming:

$$empty \equiv \neg \bigcirc true$$

$$fin(p) \equiv \Box(empty \Rightarrow p)$$

$$if\ C\ then\ P$$
$$else\ Q \equiv (C \Rightarrow P) \wedge (\neg C \Rightarrow Q)$$

$$for\ n\ times\ do\ P \equiv if\ n = 0\ then\ empty$$
$$else\ [p;\ for\ n - 1\ times\ do\ p]$$

In general, Tempura constructs intervals imperatively, rather than declaratively. It provides various advantages over the execution of discrete temporal logics, such as the ability to represent 'real-time' and to sequence operations (via ';'). However, Tempura executes a restricted version of ITL, one that is essentially deterministic. In particular, the '$\Diamond$' and '$\vee$' operators are not defined in Tempura.

Thus, a formula φ of ITL is a Tempura program if, after eliminating quantifiers and expanding predicates, it can be written in the canonical form [45], i.e.,

$$\varphi \equiv \bigwedge_{i=0}^{n} \bigcirc^i p_i$$

where p_i is a conjunction of assignments to basic propositions.

5.4   Interval Temporal Logic Programming — Tokio

Tokio [41, 51] is a logic programming language based on the extension of Prolog with ITL formulae. It provides a powerful system in which a range of applications can be implemented and verified. While Tempura executes a deterministic subset of ITL, Tokio executes an extended subset incorporating the non-deterministic operators $\Diamond$ and $\vee$.

Since Tokio is an extension of Prolog, if no temporal statements are given, the execution of Tokio reduces to that of Prolog. When ITL statements *are* present, the execution mechanism is extended to construct intervals satisfying those temporal constraints. Rather than giving a detailed description of the execution mechanism, we simply give a few sample goals, outlining the output produced within Tokio [41].

Tokio Examples

Goal: $\leftarrow write(1)$
Result: 1

Goal: $\leftarrow \bigcirc write(2), write(1)$
Result: 12

Goal: $\leftarrow length(5), \Box write(1)$
Result: 11111

Goal: $\leftarrow length(8), (length(5), \Box write(0); write(1))$
Result: 000001111

5.5   A Brief Survey of Other Approaches to Executable Temporal Logics

Temporal Prolog (Gabbay) Gabbay [44] introduces an alternative extension of the logic programming paradigm to temporal logics. Apart from considering temporal logics that incorporate the past, his principal difference to TEMPLOG is that he uses a subset of temporal Horn Clauses where only $\Diamond p$, $\bigcirc p$ and p (p being a proposition) are allowed in either the head or the body of a rule. Gabbay provides a semantics and various technical results relating to his system, but does not describe the implementation in any detail. However, he does extend the framework to a variety of different models of time and collections of temporal operators.

Executing Temporal Formulae Efficiently Merz [54] discusses the issues involved in the tradeoff between efficiency and expressiveness in the execution of temporal formulae. In particular, he defines a class of temporal logic formulae, which is both sufficiently abstract to support high-level descriptions of algorithms and yet, whose model construction problem is tractable.

Concurrent METATEM Although related to METATEM, Concurrent METATEM [38], is a refinement that is particularly suited to the modelling and implementation of distributed and concurrent systems. It imposes more practical operational constraints upon this execution and places the executing processes within a concurrent object-based framework. Concurrent METATEM has been applied to problems in distributed problem-solving [34], modelling societies [32], distributed AI [33], transport simulation [27], hospital simulation [61] and protocols for cooperative action [35]. For a longer survey of potential applications, see [29].

Metric Temporal Logic Programming Brzoska [15] presents an extension to temporal logic programming of the TEMPLOG form, incorporating not only past-time operators, but also metric temporal operators. He gives the correctness of the logic programming system in his framework and shows how it can be translated into a constraint logic programming [49] system over an appropriate algebra. This approach forms the basis for the more ambitious LIMETTE (Logic programming Integrating METric Temporal Extensions) system currently under development [17, 63].

Branching Temporal Logic Programming Tang [66], rather than providing an extension of Prolog that incorporates a form of temporal logic, extends the verification method for temporal logic to incorporate logic programming. In particular, the states of a model checker for CTL (a branching-time temporal logic [21]) are extended with Prolog like statements.

Temporal Prolog (Hrycej) Hrycej [46, 47] describes a completely different *Temporal Prolog* from Gabbay's (see above), which is based upon interval temporal logic. He extends Prolog using a form of Allen's interval temporal logic [6] and applies it to temporal knowledge representation and temporal planning problems.

Annotated Constraint Logic Programming Frühwirth [40] outlines another method for implementing temporal logics as an extension to logic programming, this time by defining a *temporal annotated logic*. A clausal fragment of annotated logics can be mapped directly on to a particular form of constraint logic programming.

6 Concluding Remarks

We have introduced a variety of methods for executing modal and temporal logics. In addition to outlining some of the more common execution mechanisms, we have indicated some of the applications and references for further study.

These areas, particularly that of executable temporal logics, are expanding. Consequently, we can expect to see, in the future, both a proliferation of new techniques and languages, and the refinement and application of some of those techniques outlined here.

Acknowledgements

The authors would like to thank all the other contributors to this volume, together with Michael Wooldridge, for their comments and suggestions regarding this paper.

References

1. M. Abadi and Z. Manna. Temporal Logic Programming. *Journal of Symbolic Computation*, 8: 277–295, 1989.
2. M. Abadi and Z. Manna. Nonclausal Deduction in First-Order Temporal Logic. *ACM Journal*, 37(2):279–317, April 1990.
3. M. Abadi. The Power of Temporal Proofs. *Theoretical Computer Science*, 64:35–84, 1989.
4. J. Allen and J. Koomen. Planning using a temporal world model. In *Proc. IJCAI-83*, pages 741–747, Karlsruhe, August 1983.
5. J. Allen. Maintaining Knowledge about temporal intervals. *Comm. ACM*, 26(11):832–843, November 1983.
6. J. Allen. Towards a general theory of action and time. *Artificial Intelligence*, 23(2):123–154, 1984.
7. J. Allen and P. Hayes. A Common Sense Theory of Time. In *Proc. IJCAI-85*, pages 528–531, Los Angeles, California, August 1985.
8. J-M. Alliot, A. Herzig and M. Lima-Marques. Implementing Prolog Extensions: a Parallel Inference Machine. In *Proceedings of the International Conference on Fifth Generation Computer Systems*, ICOT, Japan, 1992.
9. P. Balbiani, A. Herzig, and M. Lima-Marques. TIM: The Toulouse Inference Machine for Non-Classical Logic Programming. *Lecture Notes in Computer Science*, 567, 1991.
10. B. Banieqbal and H. Barringer. A study of an Extended Temporal Language and a Temporal Fixed Point Calculus. Technical Report UMCS-86-10-2, Department of Computer Science, University of Manchester, November 1986.
11. H. Barringer, M. Fisher, D. Gabbay, G. Gough, and R. Owens. METATEM: A Framework for Programming in Temporal Logic. In *Proceedings of REX Workshop on Stepwise Refinement of Distributed Systems: Models, Formalisms, Correctness*, Mook, Netherlands, June 1989. (Published in *Lecture Notes in Computer Science*, volume 430, Springer Verlag).
12. H. Barringer, M. Fisher, D. Gabbay, and A. Hunter. Meta-Reasoning in Executable Temporal Logic. In J. Allen, R. Fikes, and E. Sandewall, editors, *Proceedings of the International Conference on Principles of Knowledge Representation and Reasoning (KR)*, Cambridge, Massachusetts, April 1991. Morgan Kaufmann.
13. M. Baudinet. A Simple Proof of the Completeness of Temporal Logic Programming. In L Fariñas del Cerro and M. Penttonen, editors, *Intensional Logics for Programming*. Oxford University Press, 1992.
14. M. Baudinet. Temporal Logic Programming is Complete and Expressive. In *POPL16*, Austin, Texas, January 1989. ACM.
15. C. Brzoska. Temporal Logic Programming with Metric and Past Operators. (In this volume.)
16. C. Brzoska. Temporal logic programming and its relation to constraint logic programming. In *Proceedings of International Symposium on Logic Programming (ILPS)*, San Diego, U.S.A., November 1991. (Published by MIT Press.).

17. C. Brzoska and K. Schäfer. LIMETTE: Logic programming integrating metric temporal extensions, language definition and user manual. Interner Bericht 9/93, Fak. für Informatik, Universität Karlsruhe, 1993.

18. R. Bull and K. Segerberg. Basic Modal Logic. In D. Gabbay and F. Guenthner, editors, *Handbook of Philosophical Logic, Volume II*. Reidel, 1984.

19. J. Burgess. Basic Tense Logic. In D. Gabbay and F. Guenthner, editors, *Handbook of Philosophical Logic, Volume II*. Reidel, 1984.

20. M. Cavalcanti. Solving Air-Traffic Problems with "Possible Worlds". (In this volume.)

21. E. M. Clarke and E. A. Emerson. Using Branching Time Temporal Logic to Synthesise Synchronisation Skeletons. *Science of Computer Programming*, 2:241–266, 1982.

22. T. Dean. Large-Scale Temporal Data Bases for Planning in Complex Domains. In *Proc. IJCAI-87*, pages 860–866, Milan, August 1987.

23. N. Den Haan. Investigations into the Application of Deontic Logic. (In this volume.)

24. E. A. Emerson. Temporal and Modal Logic. In J. van Leeuwen, editor, *Handbook of Theoretical Computer Science*, pages 996–1072. Elsevier, 1990.

25. L. Fariñas del Cerro. Resolution modal logics. *Logique et Analyse*, 152–172, 1985.

26. L. Fariñas del Cerro. MOLOG: A system that extends Prolog with modal logic. *New Generation Computing*, February 1986.

27. M. Finger, M. Fisher, and R. Owens. METATEM at Work: Modelling Reactive Systems Using Executable Temporal Logic. In *Sixth International Conference on Industrial and Engineering Applications of Artificial Intelligence and Expert Systems (IEA/AIE-93)*, Edinburgh, U.K., June 1993. Gordon and Breach Publishers.

28. M. Finger, P. McBrien, and R. Owens. Databases and Executable Temporal Logic. In *Proceedings of the ESPRIT Conference*, November 1991.

29. M. Fisher. A Survey of Concurrent METATEM — The Language and its Applications. In *First International Conference on Temporal Logic (ICTL)*, July 1994.

30. M. Fisher and P. Noël. Transformation and Synthesis in METATEM – Part I: Propositional METATEM. Technical Report UMCS-92-2-1, Department of Computer Science, University of Manchester, Oxford Road, Manchester M13 9PL, U.K., February 1992.

31. M. Fisher and R. Owens. From the Past to the Future: Executing Temporal Logic Programs. In *Proceedings of Logic Programming and Automated Reasoning (LPAR)*, St. Petersberg, Russia, July 1992. (Published in *Lecture Notes in Computer Science*, volume 624, Springer-Verlag).

32. M. Fisher and M. Wooldridge. A Logical Approach to the Representation of Societies of Agents. In *Proceedings of Second International Workshop on Simulating Societies (SimSoc)*, Certosa di Pontignano, Siena, Italy, July 1993.

33. M. Fisher and M. Wooldridge. Executable Temporal Logic for Distributed A.I. In *Twelfth International Workshop on Distributed A.I.*, Hidden Valley Resort, Pennsylvania, May 1993.

34. M. Fisher and M. Wooldridge. Specifying and Verifying Distributed Intelligent Systems. In *Portuguese Conference on Artificial Intelligence (EPIA)*. Springer-Verlag, October 1993.

35. M. Fisher and M. Wooldridge. Specifying and Executing Protocols for Cooperative Action. In *Proceedings of International Working Conference on Cooperating Knowledge-Based Systems (CKBS)*, Keele, June 1994.

36. M. Fisher. A Resolution Method for Temporal Logic. In *Proceedings of the Twelfth International Joint Conference on Artificial Intelligence (IJCAI)*, Sydney, Australia, August 1991. Morgan Kaufman.

37. M. Fisher. A Normal Form for First-Order Temporal Formulae. In *Proceedings of Eleventh International Conference on Automated Deduction (CADE)*, Saratoga Springs, New York, June 1992. (Published in *Lecture Notes in Computer Science*, volume 607, Springer-Verlag).

38. M. Fisher. Concurrent METATEM — A Language for Modelling Reactive Systems. In *Parallel Architectures and Languages, Europe (PARLE)*, Munich, Germany, June 1993. Springer-Verlag.
39. M. Fitting. *Proof methods for modal and intuitionistic logics*. Reidel Publishers, 1983.
40. T. Frühwirth. Temporal Logic and Annotated Constraint Logic Programming. (In this volume.)
41. M. Fujita, S. Kono, T. Tanaka, and T. Moto-oka. Tokio: Logic Programming Language based on Temporal Logic and its compilation into Prolog. In *3rd International Conference on Logic Programming*, London, July 1986. (Published in *Lecture Notes in Computer Science*, volume 225, Springer-Verlag).
42. D. Gabbay, A. Pnueli, S. Shelah, and J. Stavi. The Temporal Analysis of Fairness. In *Proceedings of the Seventh ACM Symposium on the Principles of Programming Languages*, pages 163–173, Las Vegas, Nevada, January 1980.
43. D. Gabbay. Declarative Past and Imperative Future: Executable Temporal Logic for Interactive Systems. In B. Banieqbal, H. Barringer, and A. Pnueli, editors, *Proceedings of Colloquium on Temporal Logic in Specification*, pages 402–450, Altrincham, U.K., 1987. (Published in *Lecture Notes in Computer Science*, volume 398, Springer-Verlag).
44. D. Gabbay. Modal and Temporal Logic II (A Temporal Prolog Machine). In T. Dodd, R. Owens, and S. Torrance, editors, *Logic Programming—Expanding the Horizon*. Intellect Books Ltd, 1991.
45. R. Hale and B. Moszkowski. Parallel programming in temporal logic. In *Parallel Architectures and Languages Europe (PARLE)*, Eindhoven, The Netherlands, June 1987. (Published as Lecture Notes in Computer Science, volume 259, Springer Verlag, Berlin.).
46. T. Hrycej. Temporal Prolog. In Yves Kodratoff, editor, *Proceedings of the European Conference on Artificial Intelligence*, pages 296–301. Pitman Publishing, August 1988.
47. T. Hrycej. A temporal extension of Prolog. *The Journal of Logic Programming*, 15(1 & 2):113–145, January 1993.
48. G. Hughes and M. Cresswell. *A Companion to Modal Logic*. Methuen (UP), 1984.
49. J. Jaffir and J-L. Lassez. Constraint Logic Programming. In *Proceedings of the Fourteenth ACM Symposium on the Principles of Programming Languages*, pages 111–119, Munich, West Germany, January 1987.
50. J. A. W. Kamp. *Tense Logic and the Theory of Linear Order*. PhD thesis, University of California, May 1968.
51. S. Kono. A Combination of Clausal and Non Clausal Temporal Logic Programs. (In this volume.)
52. F. Kröger. *Temporal Logic of Programs*. Springer-Verlag, 1987.
53. Z. Manna and A. Pnueli. *The Temporal Logic of Reactive and Concurrent Systems: Specification*. Springer-Verlag, New York, 1992.
54. S. Merz. Efficiently Executable Temporal Logic Programs. (In this volume.)
55. S. Merz. Decidability and Incompleteness Results for First-Order Temporal Logics of Linear Time. In *Journal of Applied Non-Classical Logics*, 2(2), 1992.
56. B. Moszkowski. Reasoning about digital circuits. PhD Thesis, Stanford University, July 1983.
57. B. Moszkowski. *Executing Temporal Logic Programs*. Cambridge University Press, Cambridge, U.K., 1986.
58. B. Moszkowski and Z. Manna. Reasoning in Interval Temporal Logic. *Lecture Notes in Computer Science*, 164, 1984.
59. M. Orgun and W. Wadge. Theory and Practice of Temporal Logic Programming. In L Fariñas del Cerro and M. Penttonen, editors, *Intensional Logics for Programming*. Oxford University Press, 1992.

60. M. Orgun and W. Wadge. Towards a unified theory of intensional logic programming. *The Journal of Logic Programming*, 13(1, 2, 3 and 4):413–440, 1992.
61. M. Reynolds. Towards First-Order Concurrent METATEM. (In this volume.)
62. A. P. Sistla and E. M. Clarke. Complexity of propositional linear temporal logics. *ACM Journal*, 32(3):733–749, July 1985.
63. K. Schäfer. Entwicklung einer temporallogischen Sprache zur Beschreibung von Abläufen in Straßenverkehrsszenen. Diplomarbeit, Universität Karlsruhe, Inst. für Logik, Komplexität und Deduktionssysteme, 1993.
64. R. Schwartz, P. Melliar-Smith and F. Vogt. An Interval-Based Temporal Logic. *Lecture Notes in Computer Science*, 164, 1984.
65. A. Szalas and L. Holenderski. Incompleteness of First-Order Temporal Logic with Until. *Theoretical Computer Science*, 57:317–325, 1988.
66. T. Tang. Temporal Logic CTL + Prolog. *Journal of Automated Reasoning*, 5:49–65, 1989.
67. J. van Bentham. Correspondence Theory. In D. Gabbay and F. Guenthner, editors, *Handbook of Philosophical Logic, Volume II*. Reidel, 1984.
68. L. Wallen. *Automated Deduction in Non-Classical Logics*. MIT Press, 1990.

Temporal Logic Programming with Metric and Past Operators

Christoph Brzoska

Abstract. Temporal logic allows us to use logic programming to specify and to program dynamically changing situations and non-terminating computations in a natural and problem oriented way. Recently so called metric or real-time temporal logics have been proposed for the specification of real-time systems, for which not only qualitative but also quantitative temporal properties are very important. In this work we investigate a subset of metric temporal Horn logic called *MTL-programs*, for which we give a translation into CLP($\mathcal{A}'$)-programs and CLP($\mathcal{A}'$)-goals over a suitable algebra $\mathcal{A}'$. We give a restriction of the CLP($\mathcal{A}'$)-derivation mechanism sufficient for the derivation of MTL-goals from MTL-programs, which admits efficient satisfiability checking of the constraints generated. Its worst case complexity is linear in the number of variables involved, contrary to general satisfiability checking of constraints over $\mathcal{A}'$ which is NP-complete. Moreover, we show that an extension of the metric temporal logic considered by the inclusion of variables within the temporal operators leads already to a temporal Horn logic which is expressively equivalent to constraint logic programs over $\mathcal{A}'$.

1   Introduction

Temporal logic is a formalism for reasoning about a changing world. Because the concept of time is built into the formalism, it has been widely used for the specification and verification of concurrent programs [MP81, Lam83] and for the description of activities where the chronological order of events is central, such as in planning or historical databases [All84, CI88, BCW92]. The idea emerged to make temporal logic directly executable for the same reason as it has been done for first-order logic in Prolog.

In classical logic programming the various semantics usually defined — operational, functional and logical [Llo84] — coincide, while temporal Horn logic is even incomplete in general [Mer90] (i.e., there is no sound and complete proof method for temporal Horn logic). The reason is the incompleteness of first-order temporal logic [Sza86, Aba89]. On the other hand, there are fragments of temporal Horn logic admitting sound and complete proof methods that are efficient enough to serve as operational semantics for programming languages, e.g. *Templog* [AM89, Bau89], Chronolog [OW92], Temporal Prolog [Gab87, Gab91], Temporal Prolog [Hry88], BNR-Prolog, Starlog [CK91], IQ-Prolog [RJC91]. In [Bau89, OW92] classical results on semantics for logic programs have been shown for the temporal languages investigated. There is a close correspondence between temporal and constraint logic programming [Brz91] allowing to carry over various results on constraint logic programming (CLP) to temporal logic programming languages without developing new techniques specific to temporal logic.

In this work we show how the connection between temporal logic programming and constraint logic programming can be used for temporal logic programming with past operators (e.g., at some previous time point or at every previous time point) and metric temporal operators (e.g., sometime within t time points). Metric temporal logic has been proposed in [KVdR83] for the specification of real-time systems, for which not only qualitative but also quantitative properties are very important, and has recently been investigated in [Koy89, AH90, HLP90].

We define a subset of metric temporal Horn logic called *Metric Temporal Logic Programs (MTL-programs)* that allows a translation into CLP($\mathcal{A}'$)-programs over a suitable algebra $\mathcal{A}'$. In general checking of constraints over $\mathcal{A}'$ for satisfiability is NP-complete and therefore the execution of one CLP($\mathcal{A}'$)-derivation step is NP-complete as well. But we can show that during derivations of MTL-goals from MTL-programs only very restricted constraints over $\mathcal{A}'$ need to be solved. We define a restricted form of the CLP($\mathcal{A}'$)-derivation mechanism called MTL-resolution that is sufficient for those derivations and give a simple method for satisfiability checking of constraints generated during MTL-derivations. This method has linear worst case complexity with respect to the number of variables involved. Moreover, we show that the inclusion of variables within metric operators, as has been proposed in [Koy89, AH90, HLP90], leads already to a temporal Horn logic that is expressively equivalent to constraint logic programs over $\mathcal{A}'$.

2   Preliminaries

We assume the reader is familiar with the theory of logic programming and constraint logic programming (cf. [Llo84, JL86a]) and introduce only briefly our notational conventions. A *signature* Σ is a triple (S, F, P) consisting of a set S of sorts, a set $F = \bigcup_{w \in S^*, s \in S} F_{w,s}$ of function symbols, and of a set $P = \bigcup_{w \in S^*} P_w$ of predicate symbols. We write $f : s_1 \ldots s_n \to s$ if $f \in F_{s_1 \ldots s_n, s}$ and $p : s_1 \ldots s_n$ if $p \in P_{s_1 \ldots s_n}$. $Var(t)$ (respectively, $Var(A)$) denotes the set of free variables occurring in t (respectively, in A). $T_\Sigma(\mathcal{V}) = \bigcup_{s \in S} T_\Sigma(\mathcal{V})_s$ denotes terms over a signature Σ and a set of variables $\mathcal{V}$, and $T_\Sigma = \bigcup_{s \in S} T_{\Sigma s}$ denotes ground terms over Σ.

A *(Σ-)substitution* is a mapping $\sigma : \mathcal{V} \mapsto T_\Sigma(\mathcal{V})$ that is the identity except for finitely many variables and that satisfies $\sigma(\mathcal{V}_s) \subseteq T_\Sigma(\mathcal{V})_s$ for all $s \in S$. We write substitutions as finite sets of variable replacements $\{x_1 \leftarrow t_1, \ldots, x_n \leftarrow t_n\}$. For a substitution σ, $dom(\sigma) = \{x | \sigma(x) \neq x\}$. An *unifier* of two terms t, t' is a substitution σ such that $\sigma(t) = \sigma(t')$; a unifier θ is called *most general (mgu)* if for any unifier σ of t, t' there exists a substitution λ such that $\sigma = \lambda \circ \theta$. We often use postfix notation for application and composition of substitutions, i.e. we write $t\sigma$ and $\sigma\lambda$ for $\sigma(t)$ and $\lambda \circ \sigma$, respectively.

The following notational conventions are adapted throughout the paper. Unless otherwise stated, $\Sigma = (S, F, P)$ denotes the signature and V the set of variables at hand; the possibly subscripted symbols $s, s', \ldots$ denote sort symbols, the possibly subscripted symbols x, y, z denote variables, the possibly subscripted symbols t, t' denote terms and p, q denote predicate symbols. $\bar{x}$ denotes a sequence $x_1, \ldots, x_n$ of objects or — but this will be clear from the context — a sum $x_1 + \ldots + x_n$. $\exists$ (respectively, $\forall$) denotes the existential closure (respectively, universal closure).

The formulae of *metric temporal logic* are built up with the usual logical connectives and the following temporal operators: $\square_t$ (always within t time points), $\diamond_t$ (sometime within t time points), $\circ$ (next) and $\bullet$ (previous), where $t \in \bar{Z} = Z \cup \{+\infty, -\infty\}$, and a number of derived operators listed figure 2.1. In the temporal first-order logic of

$$\square A \overset{\text{def}}{=} \square_{+\infty} \square_{-\infty} A \qquad \text{(unrestricted always)}$$

$$\diamond A \overset{\text{def}}{=} \diamond_{+\infty} \diamond_{-\infty} A \qquad \text{(unrestricted sometime)}$$

$$\square_+ A \overset{\text{def}}{=} \square_{+\infty} A \qquad \text{(always in the future)}$$

$$\square_- A \overset{\text{def}}{=} \square_{-\infty} A \qquad \text{(always in the past)}$$

$$\diamond_+ A \overset{\text{def}}{=} \diamond_{+\infty} A \qquad \text{(sometime in the future)}$$

$$\diamond_- A \overset{\text{def}}{=} \diamond_{-\infty} A \qquad \text{(sometime in the past)}$$

$$\square_{[c_1, c_2]} A \overset{\text{def}}{=} \square_{\max(c_1, c_2) - \min(c_1, c_2)} \circ^{\min(c_1, c_2)} A, \qquad \text{(always between } c_1 \text{ and } c_2)$$
$$\text{where } c_i \in Z, (i = 1, 2) \text{ and } \circ^i \text{ with}$$
$$i \leq 0 \text{ denotes } \bullet^{|i|}.$$

$$\diamond_{[c_1, c_2]} A \overset{\text{def}}{=} \diamond_{\max(c_1, c_2) - \min(c_1, c_2)} \circ^{\min(c_1, c_2)} A, \qquad \text{(sometime between } c_1 \text{ and } c_2)$$
$$\text{where } c_i \in Z, (i = 1, 2) \text{ and } \circ^i \text{ with}$$
$$i \leq 0 \text{ denotes } \bullet^{|i|}.$$

Fig. 2.1. Derived operators.

interest, function symbols are considered to be time-independent; they are said to be *rigid*. However, the interpretation of predicate symbols can vary with time; they are said to be *flexible*. The formulae are interpreted over structures that we call *(MTL-)Σ-structures*.

Definition 2.1 *A (MTL-)Σ-structure $\mathcal{M}$ is a Kripke structure $(\mathcal{D}, \mathcal{T}, t_0, <, \mathcal{I})$ satisfying the following:*

- $(\mathcal{T}, t_0, <) \simeq (\mathcal{Z}, 0, <)$, *i.e. the set of time points is isomorphic to the set of integers, the first time point is $0 \in \mathcal{Z}$ and the before-relation on $\mathcal{T}$ is $<$ on $\mathcal{Z}$,*
- $\mathcal{D} = \bigcup_{s \in S} \mathcal{D}_s$ *with $\mathcal{D}_s \neq \emptyset$ for every $s \in S$,*
- *The interpretation $\mathcal{I}$ assigns to each function symbol $f : s_1 \ldots s_n \to s$ in Σ a mapping $f^{\mathcal{M}}$ from $\mathcal{D}_{s_1} \times \ldots \times \mathcal{D}_{s_n}$ to $\mathcal{D}_s$ and to each predicate symbol $p : s_1 \ldots s_n$ in Σ and each time point t in $\mathcal{T}$ a relation $p_t^{\mathcal{M}} \subseteq \mathcal{D}_{s_1} \times \ldots \times \mathcal{D}_{s_n}$.*

A *variable assignment* (into a MTL-structure $\mathcal{M}$) is a mapping $\alpha : \mathcal{V} \to \mathcal{D}$ satisfying $\alpha(\mathcal{V}_s) \subseteq \mathcal{D}_s$ for every sort symbol $s \in S$. Every variable assignment α can be extended uniquely to a homomorphism $\overline{\alpha} : \mathcal{T}_\Sigma(\mathcal{V}) \mapsto \mathcal{M}$. The *value of a term t in $\mathcal{M}$ under α* is defined by $\overline{\alpha}(t)$.

Definition 2.2 *The validity of a MTL formula A in a (MTL-)structure $\mathcal{M}$ at time point t under a variable assignment α, denoted $(\mathcal{M}, \alpha) \models_t A$, is inductively defined as follows:*

1. $(\mathcal{M}, \alpha) \models_t p(t_1, \ldots, t_n)$ *iff $p_t^{\mathcal{M}}(\overline{\alpha}(t_1), \ldots, \overline{\alpha}(t_n))$ holds in $\mathcal{M}$, for every predicate symbol $p : s_1 \ldots s_n$ in Σ and terms $t_i \in \mathcal{T}_\Sigma(\mathcal{V})_{s_i}$ $(i = 1, \ldots, n)$,*
2. $(\mathcal{M}, \alpha) \models_t \circ A$ *iff $(\mathcal{M}, \alpha) \models_{t+1} A$,*
3. $(\mathcal{M}, \alpha) \models_t \bullet A$ *iff $(\mathcal{M}, \alpha) \models_{t-1} A$,*
4. $(\mathcal{M}, \alpha) \models_t \Diamond_c A$ *iff $(\mathcal{M}, \alpha) \models_{t'} A$ for some t' such that*
   (a) $t \leq t' \leq t + c$ if $c \geq 0$ and
   (b) $t + c \leq t' \leq t$ if $c \leq 0$,
   where $t + (+\infty) = +\infty$, $t + (-\infty) = -\infty$ and $+\infty \geq c$, $-\infty \leq c$ for all $c \in \mathcal{Z}$.
5. $(\mathcal{M}, \alpha) \models_t \Box_c A$ *iff $(\mathcal{M}, \alpha) \models_{t'} A$ for all t' such that*
   (a) $t \leq t' \leq t + c$ if $c \geq 0$ and
   (b) $t + c \leq t' \leq t$ if $c \leq 0$,
   where $t + (+\infty) = +\infty$, $t + (-\infty) = -\infty$ and $+\infty \geq c$, $-\infty \leq c$ for all $c \in \mathcal{Z}$.

The validity of formulae of the form $A \wedge B, A \vee B, A \to B, \neg A, \forall x A, \exists x A$ at a time point t is defined in the usual way. A formula A *is valid (under α) in $\mathcal{M}$* iff $(\mathcal{M}, \alpha) \models_{t_0} A$; A *is valid (in $\mathcal{M}$)* iff $(\mathcal{M}, \alpha) \models_{t_0} A$ for all variable assignments $\alpha : \mathcal{V} \to \mathcal{M}$. The *satisfiability* and the *logical consequence*, denoted by $\models$, are defined as usual.

3   MTL-programs

We define *MTL-goals G* and *MTL-clauses D* by

$$G ::= \varepsilon \mid A \mid \circ G \mid \bullet G \mid \Diamond_c G,$$

and

$$D ::= A \mid \circ D \mid \bullet D \mid \Box_c D \mid D \leftarrow G.$$

ε denotes the empty goal, A denotes an atomic formula, and $c \in \mathcal{Z}$. Given a clause $A \leftarrow B$, we call A the *head* and B the *body* of $A \leftarrow B$.

Example 3.1 *Time outs in communication protocols:*

$$\Box \, (\Box_{+\infty} \, served\_in\_time(Address, Message) \leftarrow$$
$$\Diamond_{-t} \, send(Address, Message) \wedge$$
$$acknowledge(Address, Message)),$$
$$\Box \, (\Box_{+\infty} \, time\_out(Address, Message) \leftarrow$$
$$\bullet^t send(Address, Message) \wedge$$
$$not \Diamond_{-t} \, acknowledge(Address, Message)),$$

where send(a, m) models sending message m at address a, acknowledge(a, m) models getting an acknowledgement for sent message m at a. served_in_time(a, m), time_out(a, m) specifies serving request in time (respectively, not in time), and "not" denotes negation by failure.

Example 3.2 *Historical Databases:*

$$\Box_{[-3000,-2100]} \, salary(john, 15000).$$
$$\Box_{[-2099,-1100]} \, salary(john, 20000).$$
$$\Box_{[-1099,-1]} \, salary(john, 25000).$$

$$\Box_{[-3000,-2100]} \, department(john, toys).$$
$$\Box_{[-2099,-1100]} \, department(john, shoes).$$
$$\Box_{[-1099,-1]} \, department(john, clothes).$$

$$\Box_{[-3000,-1]} \, manager(toys, peter).$$
$$\Box_{[-3000,-1]} \, manager(shoes, bob).$$
$$\Box_{[-3000,-1]} \, manager(clothes, martin).$$

$$\Box \, (increase(E, M) \leftarrow \bullet salary(E, M_1) \wedge salary(E, M_2) \wedge M = M_2 - M_1).$$

The queries

1. *Who was the manager of John as he became an increase?*
2. *Has John become an increase last year?*

corresponds to MTL-goals

$$\leftarrow \Diamond_- \, (increase(john, I) \wedge department(john, D) \wedge manager(D, M) \wedge I > 0)$$
and
$$\leftarrow \Diamond_{-365} \, increase(john, Y) \wedge Y > 0.$$

Example 3.3 *Robot motion planning: Assume we have for each action a_i of the robot a minimal execution time c_i^- and a maximal execution time c_i^+, and for each pair a, b of actions a minimal and a maximal reconfiguration time ab^- and ab^+. We model the minimal and maximal execution time for each a_i by clauses*

$$\Box \, (\Box_{[c_i^-, c_i^+]} \, end(a_i) \leftarrow start(a_i))$$

and the reconfiguration time for each pair of action a and b by clauses

$$\Box \, (\Box_{[ab^-, ab^+]} \, start(b) \leftarrow end(a)).$$

The requirement of performing some actions with priority if their execution can be performed within 60 seconds can be specified using a MTL-clause, namely by

$$\Box \, (do\_next(X) \leftarrow request\_for(X) \wedge priority(X) \wedge \Diamond_{60} \, start(X)).$$

4 Translation into first-order logic

Similarly to theorem proving in modal logics by translating into first-order logic [Ohl88, Ohl89, dCH90, AE89], we will translate MTL-programs P into classical logic programs $\Pi(P)$ and MTL-goals $\leftarrow B$ into goals $\leftarrow \Pi(B)$ such that $P \models B$ iff there is a successful derivation of $\leftarrow \Pi(B)$ from $\Pi(P)$. The idea is to add an additional argument to each predicate in P and to express the temporal relations which are expressed by temporal operators in temporal logic by equations and inequalities between those arguments. Thus, the translated programs are programs over an enriched signature $\Pi(\Sigma) = (\Pi(S), \Pi(F), \Pi(P))$ which contains some additional function and predicate symbols. More precisely, the translation of a signature consists of

- $\Pi(S) = S \uplus \{int\}$,
- $\Pi(F) = F \uplus \{0 : \rightarrow int, s : int \rightarrow int, p : int \rightarrow int, + : int \, int \rightarrow int\}$ and
- $\Pi(P) = \{p : int \, s_1 \dots s_n | p : s_1 \dots s_n \in P\} \uplus \{= : int \, int, \leq : int \, int\}$,

where $\uplus$ denotes the disjoint union of sets.

The translation function Π that transforms MTL-clauses (respectively, MTL-goals) over a given signature Σ into Horn clauses (respectively, goals) over the signature $\Pi(\Sigma)$ is defined as follows:

$\Pi(A) = \pi(A, 0, \emptyset)$, where A denotes a MTL-formula and π is defined by

$\pi(\circ A, t, E) = \pi(A, s(t), E)$

$\pi(\bullet A, t, E) = \pi(A, p(t), E)$

$\pi(\Box \, A, t, E) = \pi(A, t + x, E)$, where x is a new variable of sort int, i.e. $x \notin Var(t)$.

$\pi(\Diamond \, A, t, E) = \pi(A, t + x, E)$, where x is a new variable of sort int, i.e. $x \notin Var(t)$.

$\pi(\Box_c \, A, t, E) = \begin{cases} \pi(A, t + x, E \cup \{0 \leq x \leq c\}) & \text{if } c \geq 0 \\ \pi(A, t + x, E \cup \{c \leq x \leq 0\}) & \text{if } c \leq 0, \end{cases}$

where x is a new variable of sort int.

$\pi(\Diamond_c \, A, t, E) = \begin{cases} \pi(A, t + x, E \cup \{0 \leq x \leq c\}) & \text{if } c \geq 0 \\ \pi(A, t + x, E \cup \{c \leq x \leq 0\}) & \text{if } c \leq 0, \end{cases}$

where x is a new variable of sort int.

$\pi(A \wedge B, t, E) = \pi(A, t, E) \wedge \pi(B, t, E)$

$\pi(A \leftarrow B, t, E) = \begin{cases} (p(t, \bar{t}) \leftarrow E') \leftarrow \pi(B, t, E) & \text{if } \pi(A, t, E) = p(t, \bar{t}) \wedge E' \\ (A' \leftarrow B') \leftarrow \pi(B, t, E) & \text{if } \pi(A, t, E) = A' \leftarrow B' \end{cases}$

$\pi(p(\bar{t}), t, E) = p(t, \bar{t}) \wedge E$

Remarks

1. $x \leq +\infty$ (respectively, $-\infty \leq x$) represents an empty constraint (i.e., $x \leq +\infty$ (respectively, $-\infty \leq x$) is always true).
2. We can handle the operators $\square_c$ and $\diamond_c$ in the same way, since $\diamond_c$-operators occur only in MTL-bodies and $\square_c$-operators occur only in MTL-heads and at the top of MTL-clauses.
3. For notational convenience, we often drop the last argument of π and write π as a binary function if the last argument of π is not important or it is clear from the context (i.e., we write $\pi(A, t)$ instead of $\pi(A, t, E)$).
4. In the following we write $t \pm c$ instead of $s^c(t)$ (respectively, $p^c(t)$) for convenience.

Given a MTL-formula A containing an atom $p(t_1, \ldots, t_n)$ and its translation $p(t, t_1, \ldots, t_n)$ in $\Pi(A)$, we call the term t constructed during the translation of A the *temporal context of* $p(t, t_1, \ldots, t_n)$ *in* $\Pi(A)$. We denote the set of temporal contexts of all atoms in a translated MTL-formula A by $TC(A)$ and call them *the temporal contexts of* A.

We interpret the translated MTL-formulae in classical first-order structures, called *MTL-Π-structures*, in which the denotation of *int*, 0, s, p, $+$, $=$ and $\leq$ is the set of integers $\mathcal{Z}$, the number $0 \in \mathcal{Z}$, the successor function, the predecessor function, the addition, the equality and inequality in $\mathcal{Z}$, respectively.

Definition 4.1 *Let* $\Sigma = (S, F, P)$ *be a signature,* $\mathcal{M}$ *a MTL-Σ-structure and*

$$\Pi(\mathcal{M}) = ((\mathcal{D}_s^{\Pi(\mathcal{M})})_{s \in \Pi(S)}, (f^{\Pi(\mathcal{M})})_{f \in \Pi(F)}, (p^{\Pi(\mathcal{M})})_{p \in \Pi(P)})$$

a $\Pi(\Sigma)$*-structure such that*

- $\mathcal{D}_s^{\Pi(\mathcal{M})} = \mathcal{D}_s^{\mathcal{M}}$ *for all sort symbols s in* Σ,
- $f^{\Pi(\mathcal{M})} = f^{\mathcal{M}}$ *for all function symbols f in* Σ *and*
- $p^{\Pi(\mathcal{M})}(t, d_1, \ldots, d_n)$ *holds iff* $p_t^{\mathcal{M}}(d_1, \ldots, d_n)$ *holds, for all predicate symbols p in* Σ.

The structures $\mathcal{M}$ *and* $\Pi(\mathcal{M})$ *are called corresponding structures or are said to correspond.*

Since for every MTL-structure $\mathcal{M}$ the corresponding structure $\Pi(\mathcal{M})$ is uniquely determined and vice versa, we can define a bijective function mapping MTL-structures into the corresponding Π-structures. Let us denote this function by Π and its inverse by Π^{-1}.

Lemma 4.2 *Given corresponding structures* $\mathcal{M}$ *and* $\Pi(\mathcal{M})$, *the following holds for all MTL-clauses C and all goals* $\leftarrow B$:

1. $\mathcal{M} \models C$ *iff* $\Pi(\mathcal{M}) \models \Pi(C)$,
2. $\mathcal{M} \models B$ *iff* $\Pi(\mathcal{M}) \models_{\bar{\exists}} \Pi(B)$, *where* $\bar{\exists}$ *denotes the existential closure over all variables of sort int in* $\Pi(B)$.

The question whether a MTL-goal $\leftarrow B$ is a logical consequence of a program P can now be reduced to the question of validity of $\Pi(B)$ in MTL-Π-structures.

Proposition 4.3 *Given a MTL-program P and a MTL-goal* $\leftarrow B$.

$$P \models B \text{ iff } \Pi(P) \models_{\Pi \bar{\exists}} \Pi(B),$$

where $\bar{\exists}$ *denotes the existential closure over all int-variables in* $\Pi(B)$ *and* $\models_{\Pi}$ *denotes the validity in all MTL-Π-structures.*

5 MTL-resolution

The translation of MTL-programs into a set of first-order (Horn) formulae reduces the problem whether $P \models B$ for a given MTL-program P and a MTL-goal B to the question whether $\Pi(P) \models_\Pi \Pi(B)$. The only difference between Π-structures and arbitrary structures is the fixed interpretation of the sort *int* and of the operations and predicates on this sort. The translated programs are therefore not Horn programs over the free term algebra with respect to the signature at hand, but Horn programs over the algebra

$$\mathcal{A} = (\mathcal{Z}, T_\Sigma, 0, +1, -1, +, (f)_{f \in \Sigma}, =_{\mathcal{Z}}, \leq_{\mathcal{Z}}, =),$$

where $(\mathcal{Z}, 0, +1, -1, +, =_{\mathcal{Z}}, \leq_{\mathcal{Z}})$ denotes the integers with zero, the successor function, predecessor function, the addition, the equality and inequality on $\mathcal{Z}$ and $(T_\Sigma, (f)_{f \in \Sigma}, =)$ the ground term Σ-algebra. Such Horn programs over given algebras have been proposed [JL86a] in order to enhance the expressiveness of logic programs. It has been shown [JL86b] that almost all of the results for logic programs can be generalized to Horn logic programs over arbitrary algebras if they are *solution compact* and have *corresponding first-order theories* that are *satisfaction complete*.

The algebra $\mathcal{A}$ is both solution compact and has a corresponding first-order theory that is satisfaction complete. Thus, $\Pi(\Sigma)$-programs can be considered as CLP-programs over the algebra $\mathcal{A}$ and consequently the question whether $\Pi(P) \models_\Pi \Pi(B)$ can be reduced to the question whether there exists a $(\Pi(P), \mathcal{A})$-derivation of $\Pi(B)$ from $\Pi(P)$. The solvability of equations and inequalities over $\mathcal{A}$ is decidable, since solvability of inequalities over the integers and the unifiability of terms is decidable.[1] We can therefore use $(\Pi(P), \mathcal{A})$-derivations as a sound and complete procedure for derivations of MTL-goals from MTL-programs, and all of the results for CLP-programs and CLP-derivations hold automatically for $\Pi(P)$-programs and $(\Pi(P), \mathcal{A})$-derivations.

On the other hand, solving of systems of linear equations and inequalities over the integers is well-known as a NP-complete problem. Therefore, the question arises whether the systems of equations and inequalities generated during derivations of translated MTL-goals admit more efficient methods for satisfiability checking.

In the following we will show that in fact the systems of equations and inequalities to be considered during derivations of MTL-goals are of very restricted form and that they admit very efficient methods for satisfiability checking. Essentially, the reason for this is the tree structure of equations and inequalities describing temporal contexts of translated MTL-goals and MTL-clauses.

Proposition 5.1 (cf. [Brz93b]) *Every MTL-formula containing only universal temporal operators at the top (i.e., only operators $\Box_c$, $\circ$, $\bullet$ before the first occurrence of a logical connective) can be equivalently rewritten into a formula of the form*

$$A, \Box A, \circ^i A, \bullet^i A, \Box_c \circ^i A \text{ or } \Box_c \bullet^i A$$

such that A does not contain any temporal operators at the top, $c \geq 0$ or $c \in \{+\infty, -\infty\}$ and $i \geq 0$.

---

[1] For example, a decision procedure for solving inequalities over the integers is described in [Käu88].

$$\square_{c_1} \square_{c_2} A \leftrightarrow \square_{c_1+c_2} A \qquad\qquad \text{for all } c_1, c_2 \in \mathbb{Z}$$
$$\square_+ \square_c A \leftrightarrow \square_+ A \qquad \square_c \square_+ A \leftrightarrow \square_+ A \quad \text{if } 0 \le c \in \mathbb{Z} \text{ or } c = +\infty$$
$$\square_- \square_c A \leftrightarrow \square_- \circ^c A \qquad \square_c \square_- A \leftrightarrow \square_- \circ^c A \quad \text{if } 0 \le c \in \mathbb{Z}$$
$$\square_- \square_- A \leftrightarrow \square_- A$$
$$\square_- \square_+ A \leftrightarrow \square A \qquad \square_+ \square_- A \leftrightarrow \square A$$
$$\circ \square_c A \leftrightarrow \square_c \circ A$$
$$\bullet \square_c A \leftrightarrow \square_c \bullet A$$
$$\square_c A \leftrightarrow \square_{|c|} \bullet^c A \qquad\qquad\qquad \text{if } c \le 0$$
$$\bullet \circ A \leftrightarrow A \qquad \circ \bullet A \leftrightarrow A$$
$$\square \circ A \leftrightarrow \square A \qquad \square \bullet A \leftrightarrow \square A$$
$$\square_c \square A \leftrightarrow \square A \qquad \square \square_c A \leftrightarrow \square A$$
$$\circ \diamond_c A \leftrightarrow \diamond_c \circ A \qquad \bullet \diamond_c A \leftrightarrow \diamond_c \bullet A \quad \text{for all } c \in \mathbb{Z} \cup \{+\infty, -\infty\}$$

Fig. 2.2. Equivalences of metric temporal logic.

Because of the proposition 5.1 it is sufficient to consider MTL-clauses of the form

$$\square_{[d_1^-, d_1^+]} (\square_{[d_2^-, d_2^+]} (\dots \square_{[d_{n+1}^-, d_{n+1}^+]} A \leftarrow B_n \dots B_2) \leftarrow B_1). \tag{2.1}$$

Their translation leads then to

$$(\dots (((\pi(A, x_1 + \dots x_n + x_{n+1} + b) \leftarrow C_{n+1})$$
$$\leftarrow C_n \wedge \pi(B_n, x_1 + x_2 + \dots + x_n))$$
$$\vdots$$
$$\leftarrow C_1 \wedge \pi(B_1, x_1) \dots)$$

with $b \in \mathbb{Z}$. Such clauses are logically equivalent to

$$\pi(A, \sum_{i=1}^{n+1} x_i) \leftarrow \bigcup_{i=0}^{n+1} C_i \wedge \bigwedge_{i=1}^{n} \pi(B_i, \sum_{j=1}^{i} x_j). \tag{2.2}$$

If some of the $\square_{[d_i^-, d_i^+]}$ does not occur in (2.1) then (2.2) has the form

$$\pi(A, \sum_{i=1}^{m} x_{k_i}) \leftarrow \bigcup_{i=0}^{n+1} C_i \wedge \bigwedge_{i=1}^{n} \pi(B_i, \sum_{j=1}^{\max\{l|k_l \le i\}} x_{k_j}), \tag{2.3}$$

where $k_1 < k_2 < \dots < k_m \le n + 1$, which is the general form of the translated MTL-clauses.[2] Moreover, translation of goals that are normalized with respect to the equivalences given in figure 2.2 yields $(\Pi(P), \mathcal{A})$-goals that have the form

$$\leftarrow C \wedge \pi(A, \bar{x} + y + c) \wedge \pi(B, \bar{x} + y) \wedge \bigwedge_{i=0}^{n} \pi(B_i, \bar{x}_{n-i}), \tag{2.4}$$

[2] Because of $\square_{[0,0]} A \leftrightarrow A$ we can transform each MTL-clause into a clause of the form given in (2.1) by introduction of new redundant operators. But this results into superfluous variables in the translated clauses and thereby into less efficient CLP-programs.

$$\frac{\pi(A', \sum_{i=1}^{m} x_{k_i} + b) \leftarrow C' \cup \bigcup_{i=1}^{m} \{c_{k_i}^- \leq x_{k_i} \leq c_{k_i}^+\} \atop \wedge \bigwedge_{i=1}^{n} \pi(B'_i, \sum_{j=1}^{\max\{l | k_l \leq i\}} x_{k_j})}{\leftarrow C \wedge \pi(A, \bar{y} + c) \wedge \bigwedge_{i=0}^{n} \pi(B_i, \bar{y}_{n-i})}$$

$$\frac{}{\leftarrow (C \cup C' \cup \{c_{k_1}^- \leq \bar{y} + \sum_{i=m}^{2} -x_{k_i} + c - b \leq c_{k_1}^+\} \cup \bigcup_{i=2}^{m} \{-c_{k_i}^+ \leq -x_{k_i} \leq -c_{k_i}^-\}} \tag{2.5}$$
$$\wedge \bigwedge_{i=1}^{n} \pi(B'_i, \bar{y} + \sum_{j=m}^{\min\{l | k_l > i\}} x_{k_j} + c - b) \wedge \bigwedge_{i=0}^{n} \pi(B_i, \bar{y}_{n-i}))\theta,$$

where θ is the mgu of A and A'.

$$\frac{\pi(A', c') \leftarrow C' \wedge \pi(B', 0)}{\leftarrow C \cup \{c^- \leq y \leq c^+\} \wedge \pi(A, \bar{x} + y + c) \wedge \pi(B, \bar{x} + y) \wedge \bigwedge_{i=0}^{n} \pi(B_i, \bar{x}_{n-i})}$$

$$\frac{}{\leftarrow (C' \cup C \cup \{-c^+ - c + c' \leq \bar{x} \leq -c^- - c + c'\}} \tag{2.6}$$
$$\wedge \pi(B', 0) \wedge \pi(B, -c + c') \wedge \bigwedge_{i=0}^{n} \pi(B_i, \bar{x}_{n-i}))\theta$$

where θ is the mgu of A and A'

$$\frac{\pi(A', c') \leftarrow C' \wedge \pi(B', 0)}{\leftarrow C \wedge \pi(A, c) \wedge \pi(B, 0)}$$

$$\frac{}{\leftarrow (C' \cup C \wedge \pi(B', 0) \wedge \pi(B, 0))\theta} \text{ if } c' = c, \text{ where } \theta \text{ is the mgu of } A \text{ and } A' \tag{2.7}$$

Fig. 2.3. MTL-resolution rules.

where $c \in \mathcal{Z}$, $\bar{x} = x_1 + \ldots + x_n$, $\bar{x}_i = x_1 + \ldots + x_i$, A is an atom, C is a set of inequalities and B, B_i are bodies ($i = 0, \ldots, n$).

The $(\Pi(P), \mathcal{A})$-derivation steps needed for the derivation of translated MTL-goals from translated MTL-programs simplify therefore to the $(\Pi(P), \mathcal{A})$-derivation steps listed in the figure 2.3. In the following we call this restricted form of the $(\Pi(P), \mathcal{A})$-derivation mechanism *MTL-resolution*. MTL-resolution forms however only one part of the $(\Pi(P), \mathcal{A})$-derivation mechanism, since the sets of inequalities generated by applications of the rules of the MTL-resolution have to be checked for satisfiability. While the sets of inequalities obtained as result of the translation of MTL-clauses and MTL-goals are always satisfiable due to the syntax of MTL-formulae, the sets derived by MTL-resolution may be unsatisfiable. We will characterize these sets in order to develop efficient methods for satisfiability checking. The main idea is that the tree structure of MTL-formulae is found again in equations and inequalities describing temporal contexts of translated MTL-goals and MTL-clauses, and that this structure is kept under applications of MTL-resolution rules.

Theorem 5.2 (Soundness, completeness) *Let P be a MTL-program and G a MTL-goal. Then the following holds:*

(Completeness) *If $\leftarrow G \vdash_{(\Pi(P), \mathcal{A})} \leftarrow G'$, then $\leftarrow G \vdash_{MTL} \leftarrow G'$,*
(Soundness) *If $\leftarrow G \vdash_{MTL} \leftarrow G'$, then $\leftarrow G \vdash_{(\Pi(P), \mathcal{A})} \leftarrow G'$,*

where $\vdash_{MTL}$ stands for derivable using MTL-resolution rules.

Proof: The MTL-resolution rules (2.5–2.6) constitute a case analysis on the form of the translated MTL-clauses (cf. clause 2.3) considering goals in which atomic formulae are in scope of at least one $\Diamond_c$-operator. The rule (2.7) treats the case of an atom that is not in scope of any $\Diamond_c$-operator. Notice that the substitution for x_{k_1} in rule (2.5) (respectively, for y in the rule (2.6)) keeps the form of the translated goal as described in (2.4). Therefore, it is sufficient to prove that the MTL-rules are equivalent to the corresponding $(\Pi(P), \mathcal{A})$-derivation steps.

Rule (2.5): The $(\Pi(P), \mathcal{A})$-derivation step yields

$$\leftarrow (C \cup C' \cup \bigcup_{i=1}^{m} \{c_{k_i}^- \leq x_{k_i} \leq c_{k_i}^+\} \cup \{\underbrace{\sum_{i=1}^{m} x_{k_i} + b = \bar{y} + c}_{e}\} \wedge$$

$$\bigwedge_{i=1}^{n'} \pi(B'_i, \overset{\max\{l|k_l \leq i\}}{\sum_{j=1}} x_{k_j}) \wedge \bigwedge_{i=0}^{n} \pi(B_i, \bar{y}_{n-i}))\theta.$$

By rewriting the equation e with x_{k_1} as subject we get the equation $x_{k_1} = \bar{y} - \sum_{i=2}^{m} x_{k_i} + c - b$. Since x_{k_1} doesn't occur in $C, C', \bigcup_{i=2}^{m} \{c_{k_i}^- \leq x_{k_i} \leq c_{k_i}^+\}, \bigwedge_{i=0}^{n} \pi(B_i, \bar{y}_{n-i})$, we get by elimination of x_{k_1} the goal

$$\leftarrow (C \cup C' \cup \{c_{k_1}^- \leq \bar{y} + \sum_{i=2}^{m} -x_{k_i} + c - b \leq c_{k_1}^+\} \cup \bigcup_{i=2}^{m} \{c_{k_i}^- \leq x_{k_i} \leq c_{k_i}^+\} \wedge$$

$$\bigwedge_{i=1}^{n'} \pi(B'_i, \underbrace{\overset{\max\{l|k_l \leq i\}}{\sum_{j=2}} (x_{k_j}) - \sum_{j=2}^{m} x_{k_j} + c - b}_{\sum_{j=m}^{\min\{l|k_l > i\}} -x_{k_j} + c - b}) \wedge \bigwedge_{i=0}^{n} \pi(B_i, \bar{y}_{n-i}))\theta,$$

which is equivalent to the conclusion of the MTL-resolution rule (2.5).

Rule (2.6–2.7): We can show by similar arguments as in the foregoing case that the conclusions of the MTL-rules are equivalent to the corresponding $(\Pi(P), \mathcal{A})$-derivation steps.

$\square$

Definition 5.3 (Tree, unique prefix property) *Let T be a set of terms of the form $\sum x_i + c$ such that x_i are pairwise distinct variables and $c \in \mathcal{Z}$.[3]*

We say that T defines a tree iff for every variable x in $Var(T^*)$ there exists an unique prefix $\sum_{i=1}^{n} x_i$ in T^* with $x_n = x$ (unique prefix property), where T^* denotes the prefix closure of T.[4]

---

[3] Terms $x_1 + \ldots x_n + c$ are interpreted here as words over $\mathcal{V} \cup \mathcal{Z}$ with + as concatenation on words.

[4] A similar notion called *prefix-stability* was introduced in the context of automated theorem proving in modal logics by translating into first-order logic [Ohl88]. This property of terms coding modal contexts of translated modal logic formulae guarantees that the unification under associativity of those terms always yields a finitely set of most general unifiers although unification under associativity is infinitary in general.

Remark: If T defines a tree, then $(Var(T), \leq)$ is a tree, where $\leq$ is definied by: $x \leq y$ iff there exist $\bar{x}, \bar{y} \in T^*$ such that $\bar{x} + x$ and $\bar{y} = \bar{x} + x + y$. This follows easily, since $t \not\leq \lambda$, where λ denotes the empty word and $t \neq \lambda$, and for every variable $x \in Var(T)$ there exists a unique prefix $x_1 + \ldots + x_n$ with $x_n = x$ (unique prefix property).

Definition 5.4 (Tree constraint system) *A set of inequalities C is called a tree constraint system (with respect to $(Var(C), \leq))$ iff*

$$C \subseteq \{c^- \leq x, x \leq c^+ | x \text{ is a variable and } c^-, c^+ \in \mathcal{Z} \} \cup$$

$$\{c_j^- \leq \sum_{i=j_1}^{j_n} x_i, \sum_{i=j_1}^{j_n} x_i \leq c_j^+ | \sum_{i=j_1}^{j_n} x_i \text{ is a node of } (Var(C), \leq) \text{ and } c_j^-, c_j^+ \in \mathcal{Z}\}.$$

Lemma 5.5 *Let $\Pi(P)$ be a set of translated MTL-clauses and $\Pi(G)$ a translated MTL-goal.*

- *Every clause $C = A \leftarrow c \wedge B$ in $\Pi(P)$ contains only tree constraint systems, i.e. c is a tree constraint system.*
- *If $\Pi(G) \vdash_{MTL} G'$ then $G'\theta_{G'}$ contains only tree constraint systems, where $\theta_{G'}(x) = -x$ if x occurs with negative sign in G' and $\theta_{G'}(x) = x$ otherwise.*

The restricted form of inequalities generated during MTL-derivations can be exploited for efficient satisfiability checking. The simplification rules listed in the figure 2.4 form a decision procedure for the satisfiability of inequalities that arise during MTL-derivations. (ILB) and (IUB) rules perform variable elimination according to the Fourier-Motzkin algorithm eliminating the variables being leaves of the tree underlying the tree constraint system of interest (cf. [Sch86, LM92]).

(ILB) $C \cup \{c_1 \leq \bar{x} + y, c_2 \leq \bar{x}, y \leq c_3\} \rightarrow_{sc} C \cup \{c_1 \leq \bar{x} + y, c_1 - c_3 \leq \bar{x}, y \leq c_3\}$
          if $c_1 \neq -\infty$, $c_3 \neq +\infty$ and $c_1 - c_3 > c_2$.

(IUB) $C \cup \{\bar{x} + y \leq c_1, \bar{x} \leq c_2, c_3 \leq y\} \rightarrow_{sc} C \cup \{\bar{x} + y \leq c_1, \bar{x} \leq c_1 - c_3, c_3 \leq y\}$
          if $c_1 \neq +\infty$, $c_3 \neq -\infty$ and $c_2 > c_1 - c_3$.

(MLB) $C \cup \{c_1 \leq \bar{x}, c_2 \leq \bar{x}\} \rightarrow_{sc} C \cup \{\max(c_1, c_2) \leq \bar{x}\}$

(MUB) $C \cup \{\bar{x} \leq c_1, \bar{x} \leq c_2\} \rightarrow_{sc} C \cup \{\bar{x} \leq \min(c_1, c_2)\}$,

where $\bar{x} = \sum x_i$, x_i, y are variables and $c_i \in \mathcal{Z}$.

Fig. 2.4. Satisfiability checking of tree constraint systems.

Theorem 5.6 ([Brz93b]) *Let C be a tree constraint system.*

(Termination) *There is no infinite chain $C = C_1 \rightarrow_{sc} C_2 \rightarrow_{sc} \cdots$*

(Invariance) *If* $C \rightarrow_{SC} C'$ *then* $[\![C]\!] = [\![C']\!]$, *where* $[\![C]\!]$ *denotes the set of satisfiers of C, i.e.*

$$[\![C]\!] = \{\alpha : \mathcal{V} \rightarrow \mathcal{Z} \mid C\alpha \text{ is valid in every } MTL\text{-}\Pi\text{-structure}\}.$$

(Completeness) *If* C *is unsatisfiable, then there exists a* C' *such that* $C \xrightarrow{*}_{SC} C'$ *and* C' *contains an inequality* $c_1 \leq \bar{x} \leq c_2$ *with* $c_1 > c_2$.

Furthermore, we can prove the following complexity result.

Theorem 5.7 *Let* P *be a MTL-program and* $G = \leftarrow C_0 \wedge B_0$ *be a MTL-goal with* C_0 *normalized with respect to* $\rightarrow_{SC}$.

If $\leftarrow C_0 \wedge B_0 \vdash_{MTL} G_1 = \leftarrow C_1 \wedge B_1$, *then the satisfiability of* C_1 *is decidable in* $O(n)$, *where n denotes the number of variables in* C_1.

Remark: In general, we can decide the satisfiability of C_1 in $O(depth(C_1))$, where $depth(C_1)$ denotes the depth of the tree underlying C_1.

Proof: The number of variables in C_1 corresponds to the number of nodes of $(Var(C_1), \leq$). The rules (MLB) and (MUB) are only applicable to C_1 if G_1 has been obtained using MTL-rule (2.6), since C_0 is normalized with respect to $\rightarrow_{SC}$. This MTL-rule introduces only two new inequalities in C_1. Therefore, the normal form of C_1 with respect to these rules can be obtained applying each of the rules (MLB) and (MUB) only once.

Since the rules (ILB) and (IUB) compute only improved bounds for strict prefixes of the applied inequalities $\bar{x} + y$, we can compute the normal form of C_1 applying rules (MLB) and (MUB) at the beginning and thereafter rules (ILB) and (IUB) in a bottom up manner, respectively. Therefore, if $depth(C_1)$ denotes the depth of the tree defined by C_1, then each of the rules (ILB) and (IUB) can be applied to C_1 only $depth(C_1)$ times, respectively.

Thus, we can compute the normal form of C_1 in linear time with respect to the number of variables in C_1. $\square$

6 Temporal operators with variables

In this section we investigate the expressive power of an extensions of MTL-programs by temporal operators including variables, i.e. operators of the form o^x, $\Diamond_x$, and $\Box_x$ for a variable x. Such operators are not considered in [Koy89] but the usage of variables within metric temporal operators is essential for the real time logics proposed in [AH89, AH90, HLP90].

Unfortunately, already the inclusion of the temporal operator o^x (holds exactly in distance x from now) in MTL-programs and MTL-goals leads to a logic that has the full expressive power of linear arithmetical constraints over the time structure considered.[5]

---

[5] In [KVdR83] the operator o^t is denoted by $\Diamond_{=t}$.

We encode a set of linear inequalities

$$a_{11}x_1 + \ldots + a_{1n_1}x_n \leq b_1$$
$$\vdots$$
$$a_{m1}x_1 + \ldots + a_{mn_m}x_n \leq b_m \tag{2.8}$$

by the temporal program $P =$

$$p \leftarrow \underbrace{\circ^{x_1} \ldots \circ^{x_1}}_{a_{11}} \ldots \underbrace{\circ^{x_n} \ldots \circ^{x_n}}_{a_{1n_1}} q_1 \wedge$$
$$\vdots$$
$$\underbrace{\circ^{x_1} \ldots \circ^{x_1}}_{a_{m1}} \ldots \underbrace{\circ^{x_n} \ldots \circ^{x_n}}_{a_{mn_m}} q_m$$

$$\Box_{b_i} \Box_- q_i \qquad\qquad (i = 1, \ldots, m).$$

p follows from P iff (2.8) is solvable since $P \models p$ iff

$$P \models \underbrace{\circ^{x_1} \ldots \circ^{x_1}}_{a_{11}} \ldots \underbrace{\circ^{x_n} \ldots \circ^{x_n}}_{a_{1n_1}} q_1 \wedge$$
$$\vdots$$
$$\underbrace{\circ^{x_1} \ldots \circ^{x_1}}_{a_{m1}} \ldots \underbrace{\circ^{x_n} \ldots \circ^{x_n}}_{a_{mn_m}} q_m, \tag{2.9}$$

and since (2.9) is valid iff there is a assignment α such that

$$\alpha(a_{11}x_1 + \ldots + a_{1n_1}x_n) \leq b_1$$
$$\vdots$$
$$\alpha(a_{m1}x_1 + \ldots + a_{mn_m}x_n) \leq b_m$$

holds because of the facts $\Box_{b_i} \Box_- q_i$ $(i = 1, \ldots, m)$ in P.

7 Conclusion and related work

We investigated a fragment of metric temporal Horn logic, for which we gave a translation into CLP($\mathcal{A}'$)-programs and CLP($\mathcal{A}'$)-goals over a suitable algebra $\mathcal{A}'$. We gave a restriction of the CLP($\mathcal{A}'$)-derivation mechanism sufficient for derivation of MTL-goals from MTL-programs, which admits efficient satisfiability checking of the constraints generated. Its worst case complexity is linear in the number of variables involved contrary to general satisfiability checking of constraints over $\mathcal{A}'$, which is NP-complete. We showed that the inclusion of variables within the temporal operators leads already to a temporal Horn logic that is expressively equivalent to constraint logic programs over $\mathcal{A}'$.

We think that the results of the paper suggest not to see temporal logic programming as only an instance of the constraint logic programming scheme but rather to develop and to adapt constraint logic programming techniques for temporal reasoning in logic programming. In a subsequent paper we will show how the notion of tree constraint systems with specialized constraint checking and manipulation techniques can be exploited for a richer temporal logic programming language admitting $\Box_c$ with $c \in \mathcal{Z}$, $\mathcal{S}$ (since), and $\mathcal{U}$ (until) operators in goals and bodies [Brz93a].

MTL-resolution and its extension described in [Brz93a] form the theoretical backbone of the LIMETTE system — Logic programming Integrating METric Temporal Extensions — developed by Karl Schäfer at the University of Karlsruhe [BS93, Sch93].

Templog, the most extensively investigated temporal logic programming language, proposed in [AM89] and investigated in [Bau89, Brz91, BCW92] is a fragment of MTL-programs that contains only $\Box_+$, $\circ$, $\diamondsuit_+$ operators. The execution mechanism of Templog, the TSLD-resolution, is based on the equivalences $\circ\diamondsuit_+A \leftrightarrow \diamondsuit_+\circ A$, $\circ(A \wedge B) \leftrightarrow \circ A \wedge \circ B$, $\diamondsuit_+(\diamondsuit_+A_1 \wedge \ldots \wedge \diamondsuit_+A_n) \leftrightarrow \diamondsuit_+A_1 \wedge \ldots \wedge \diamondsuit_+A_n$ allowing to simplify each Templog goal into a goal of the form $\diamondsuit_+(B' \wedge \circ^i A \wedge B'')$ or $\leftarrow B' \wedge \circ^i A \wedge B''$ for some atomic formula A. Such normal forms of goals however do not always exists. In fact MTL-goals can not be rewritten into goals with a bounded nesting of sometime operators as can be seen on the following formulae A_i^+, A_i^- defined by $A_0^+ = \diamondsuit_+A^+$, $A_0^- = \diamondsuit_-A^-$, $A_{n+1}^+ = \diamondsuit_+(A_n^+ \wedge A_n^-)$, and $A_{n+1}^- = \diamondsuit_-(A_n^+ \wedge A_n^-)$, where A^+, A^- are distinct atomic formulae.

The languages investigated in [Gab87, Gab91] are based on a different subset of temporal operators. He studied a variety of implication languages in the uniform framework of *labeled deduction systems*. The simplest one is very close to Horn logic while the most expressive one covers full temporal logic. Contrary to our approach, his language is based on sometime operators — $\diamondsuit_+$ and $\diamondsuit_-$ — which can occur in heads and bodies of implications. MTL-clauses of the form $\Box (A \rightarrow \Box B)$ are not allowed there as clauses but clauses of the form $\Box (B \rightarrow \diamondsuit_+A)$. He studied therefore primarily the problem of handling skolem functions introduced by $\diamondsuit_+$- and $\diamondsuit_-$-operators in the heads of implications and proof methods for dealing with them. The issues of efficiency, which is the main topic of our contribution, was not studied in the paper.

BNR-Prolog and *Starlog* [CK91] use interval arithmetics over the reals with $+$, $*$, $=$, $\leq$ to model intervals and to describe temporal properties in logic programs. In [RJC91] a subset of IQ-logic developed by B. Richards is used as a basis for a temporal logic programming language IQ-Prolog. The language uses a number of temporal operators indexed by terms specifying their temporal range. The operational semantics of the language is given by a translation into a constraint logic programming language with linear constraints over the time domain considered. Since the temporal operators include variables the complexity of satisfiability checking during CLP-derivations should be as high as the complexity of solving linear constraints (cf. section 6).

Furthermore, J. Chomicki and T. Imieliński considered a temporal extension of DATALOG (logic programming without function symbols) obtained by tagging each predicate with an additional argument modeling time. Due to the restrictions to one monadic function symbol modeling time and to constants and variables modeling data they obtained a decidable Horn logic. The expressive power of the language coincides with Templog without function symbols.

A. Porto and Critina Ribeiro [PR92] proposed an interval temporal logic *MI* for knowledge based systems described by temporal Horn clauses. They consider a language with skolem functions in data bases (i.e. programs) with partially specified temporal relations and study the problem of consistent completion of these temporal relations in order to prove a given goal.

For sake of completeness, let us mention the work of [OW92] and [BCH88] on semantics of "intensional" (modal) logic programming and on the Molog system [dC86, BHM91].

A completely different approach to *temporal logic programming* is taken in [Tan93, Mos86, Hal87, Gab89, BFG+89, FKTMo86] and in [Mer92]. Contrary to the logic programming paradigm, which sees program execution as deduction in computationally tractable fragments of suitable logics, program execution is considered there as construction of Kripke models for the program formulae. The main motivation of this line of research is to provide a logical basis for the specification, verification and execution of imperative programs [Tan93, Mos86, Hal87, Mer92], for the combination of logic and imperative programming [FKTMo86, Gab89] and for programming of *reactive* systems [Gab89, BFG+89].

Acknowledgements: I am grateful to Karl Schäfer for fruitful discussions and Michael Gollner, Christian Schulte, and Andreas Werner for critical reading of an earlier version of this paper. This work has been supported by the Deutsche Forschungsgemeinschaft as part of the SFB 314 (S 2).

References

[Aba89] M. Abadi. The power of temporal proofs. *Theoretical Computer Science*, 65:35–83, 1989.

[AE89] Y. Auffray and P. Enjalbert. Modal theorem proving: An equational viewpoint. In *Proc. of the 11th International Joint Conference on Artificial Intelligence*, volume 1, Detroit, 1989.

[AH89] R. Alur and T. A. Henzinger. A really temporal logic. Technical Report STAN-CS-89-1267, Stanford University, Dept. of Computer Science, 1989.

[AH90] R. Alur and T. A. Henzinger. Real-time logics: Complexity and expressiveness. In *Proc. of the 5th IEEE Symposium on Logic in Computer Science*, Philadelphia, 1990. IEEE Computer Society Press.

[All84] J. Allen. Towards a general theory of action and time. *Artificial Intelligence*, 23(2):123–154, 1984.

[AM89] M. Abadi and Z. Manna. Temporal logic programming. *J. Symbolic Computation*, 8:277–295, 1989.

[Bau89] M. Baudinet. Temporal logic programming is complete and expressive. In *Proceedings of the Sixteenth Annual ACM Symposium on Principles of Programming Languages*, Austin, Texas, January 1989.

[BCH88] Ph. Balbiani, L. Farinas Del Cerro, and A. Herzig. Declarative semantics for modal logic programs. In *Proceedings of the International Conference on Fifth Generation Computer Systems*, 1988.

[BCW92] M. Baudinet, J. Chomicki, and P. Wolper. Temporal deductive databases. In A. Tansel, J. Clifford, S. Gadia, S. Jajodia, A. Segev, and R. Snodgrass, editors, *Temporal Databases*. Benjamin/Cummings, 1992.

[BFG+89] H. Barringer, M. Fisher, D. Gabbay, G. Gought, and R. Owens. A framework for programming in temporal logic. In *Stepwise Refinement of Distributed Systems, Models, Formalisms, Correctness, REX Workshop*, Mook, Netherlands, 1989. LNCS 430, Springer Verlag.

[BHM91] P. Balbiani, Andreas Herzig, and Mamede Marques. TIM: The Toulouse inference machine. In H. Boley and M. M. Richter, editors, *Processing Declarative Knowledge - International Workshop PDK'91*. Springer Verlag, LNAI 567, 1991.

[Brz91] C. Brzoska. Temporal logic programming and its relation to constraint logic programming. In *Proc. of the 1991 Logic Programming Symposium*, San Diego, California, October 1991. MIT Press.

[Brz93a] C. Brzoska. Temporal logic programming with bounded universal (modality) goals. In *Proc. of the 10th International Conference on Logic Programming*, Budapest, Hungary, 1993. MIT Press.

[Brz93b] C. Brzoska. Temporal logic programming with metric and past operators based on constraint logic programming. Interner Bericht 2/93, Universität Karlsruhe, Fak. für Informatik, 1993.

[BS93] C. Brzoska and K. Schäfer. LIMETTE: Logic programming integrating metric temporal extensions, language definition and user manual. Interner Bericht 9/93, Fak. für Informatik, Universität Karlsruhe, 1993.

[CI88] J. Chomicki and T. Imieliński. Temporal deductive databases and infinite objects. In *Proc. of the 7th Symposium on Principles of Database Systems*, Austin, Texas, 1988.

[CK91] J. G. Cleary and V. Kaushik. Updates in a temporal logic programming language. Research Report 91/427/11, Univ. of Calgary, Dept. of Computer Science, 1991.

[dC86] Luis Farinas del Cerro. Molog: A system that extends PROLOG with modal logic. *New Generation Computing*, 4:35–50, 1986.

[dCH90] L. Farinas del Cerro and A. Herzig. Automated quantified modal logic. In P. Brazdil and K. Konolige, editors, *Machine Learning, Meta-Reasoning and Logics*. Kluwer Academic Publishers, 1990.

[FKTMo86] M. Fujita, S. Kono, T. Tanaka, and T. Moto-oka. Tokio: Logic programming language based on temporal logic and its compilation to prolog. In *Proc. of the 3rd International Conference on Logic Programming*, pages 695–709. Springer-Verlag, LNCS 225, 1986.

[Gab87] D. M. Gabbay. Modal and temporal logic programming. In A. Galton, editor, *Temporal Logics and Their Applications*, chapter 6, pages 197–237. Academic Press, London, December 1987.

[Gab89] D. M. Gabbay. Declarative past and imperative future. In B. Banieqbal, H. Barringer, and A. Pnueli, editors, *Proc. of Colloquium on Temporal Logic and Specification*, LNCS 398, pages 76–89, Altrincham, 1989. Springer Verlag.

[Gab91] D. M. Gabbay. A temporal logic programming machine. In T. Dodd, R. Owens, and S. Torrance, editors, *Logic Programming: Expanding the Horizonts*, chapter 3, pages 82–123. Intellect Books, 1991.

[Hal87] R. Hale. Temporal logic programming. In A. Galton, editor, *Temporal Logics and their Applications*, chapter 3, pages 91–119. Academic Press, London, December 1987.

[HLP90] E. Harel, O. Lichtenstein, and A. Pnueli. Explicit clock temporal logic. In *Proc. of the 5th Annual IEEE Symposium on Logic in Computer Science*, Philadelphia, 1990. IEEE Computer Society Press.

[Hry88] T. Hrycej. Temporal prolog. In *ECAI 88 Proceedings of the 8th European Conference on Artificial Intelligence*, Munich, August 1988.

[JL86a] J. Jaffar and J.-L. Lassez. Constraint logic programming. In *Proc. of the 14th ACM Symposium on Principles of Programming Languages*, pages 111–119, Munich, 1986.

[JL86b] J. Jaffar and J.-L. Lassez. Constraint logic programming. Technical report, Department of Computer Science, Monash University, Australia, June 1986.

[Käu88] T. Käufl. Simplification and decision of linear inequalities over the integers. Technical Report 9/88, University of Karlsruhe, Fak. für Informatik, 1988.

[Koy89] R. Koymans. *Specifying Message Passing and Time-Critical Systems with Temporal Logic*. PhD thesis, Technical University of Eindhoven, 1989.

[KVdR83] R. Koymans, J. Vytopil, and W. P. de Roever. Real-time programming and asynchronous message passing. In *Proc. of the 2nd ACM Symp. on Principles of Distributed Computing*, pages 187–197, Montreal, Canada, 1983.

[Lam83] L. Lamport. Specifying concurrent programs modules. *ACM Transactions on Programming Languages and Systems*, 5:190–222, April 1983.

[Llo84] J. W. Lloyd. *Foundations of Logic Programming*. Springer-Verlag, 1984.

[LM92] J.-L. Lassez and M. J. Maher. On fourier's algorithm for linear arithmetic constraints. *Journal of Automated Reasoning*, 9(3), December 1992.

[Mer90] S. Merz, October 1990. Private communication.

[Mer92] S. Merz. *Temporal Logic as a Programming Language*. Dissertation, Ludwig-Maximilians. Universität, München, 1992.

[Mos86] B. Moszkowski. *Executing Temporal Logic Programs*. Cambridge University Press, Cambridge, 1986.

[MP81] Z. Manna and A. Pnueli. Verification of concurrent programs: the temporal framework. In R. S. Boyer and J. S. Moore, editors, *The Correctness Problem in Computer Science*, pages 215–273. Academic Press, London, 1981.

[Ohl88] H. J. Ohlbach. A resolution calculus for modal logics. In E. Lusk and R. Overbeek, editors, *Proceedings of the 9th International Conference on Automated Deduction*. Springer-Verlag, LNCS 310, 1988.

[Ohl89] H. J. Ohlbach. Context logic. SEKI Report SR-89-08, FB Informatik, University of Kaiserslautern, 1989.

[OW92] M. A. Orgun and W. W. Wadge. Towards a unified theory of intensional logic programming. *Journal of Logic Programming*, 13(4):413 ff, 1992.

[PR92] A. Porto and C. Ribeiro. Temporal inference with a point-based interval algebra. In B. Neumann, editor, *Proc. of the ECAI'92 10th European Conference on Artificial Intelligence*, Wien, 1992. John Wiley & Sons.

[RJC91] B. Richards, Y. Jiang, and H. Choi. On interval-based temporal planning: An IQ strategy. In Z. Ras, editor, *Proc. of the 6th ISMIS'91*, LNAI 542. Springer Verlag, 1991.

[Sch86] A. Schriver. *Theory of Linear and Integer Programming*. Wiley, 1986.

[Sch93] K. Schäfer. Entwicklung einer temporallogischen Sprache zur Beschreibung von Abläufen in Straßenverkehrsszenen. Diplomarbeit, Universität Karlsruhe, Inst. für Logik, Komplexität und Deduktionssysteme, 1993.

[Sza86] A. Szalas. Concerning the semantic consequence relation in first-order temporal logic. *Theoretical Computer Science*, 47:329–334, 1986.

[Tan93] Ch.-S. Tang. Toward a unified logic basis for programming languages. In R.E.A. Mason, editor, *Proc. Information Processing 83*. Elsevier Science Publisher B.V. (North Holland), 1993.

A Combination of Clausal and Non Clausal Temporal Logic Programs

Shinji Kono

Abstract. We have developed Tokio interpreter[5] for first order Interval Temporal Logic[11] and an automatic theorem prover [6, 7] for Propositional Interval Temporal Logic. The verifier features deterministic tableau expansion and binary decision tree representation of subterms. Combining these, we can avoid repeated similar clausal form time constraints, and it is possible to execute wider range of specifications without time-backtracking.

1   Interval Temporal Logic

Interval Temporal Logic[11] (hereafter referred as ITL) uses a sequencing modal operator as its basis. In this logic, it is very easy to express control structures in conventional programming languages, (such as ';' and while statements). From this point of view a process algebra such as CCS[9] or CIRCAL[8] is also good for control structures, but it does not support negation and declarative expressions (like sometime or always) which are common to Temporal Logic.

In this paper, we show an implementation of an interval temporal logic theorem prover. This method is a tableau driven method[12, 14] and a practical implementation of [5, 7]. It also generates a deterministic state diagram as a verification result.

We have developed a first order Interval Temporal Logic interpreter using a kind of clausal form. It is easy to combine the result of verification and the interpreter, since the generated state diagram can be easily translated into clausal form.

First we show an informal visual representation of basic operators in ITL. An interval is a finite line which has number of clock ticks. An operator *empty* is true on the length 0 interval.

A local variable p means p occurs at the beginning of the interval.

The *nextoperator*, @P, means P becomes true after one clock cycle. Thus, in ITL @P's interval must be one clock cycle longer than P's and @P is false on the empty interval. We call this *strongnext*. We write *weaknext* ◯P as @P ∨ *empty*. P can be any temporal logic formula.

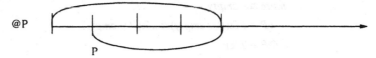

We introduce the chop operator '&' which combines two intervals. P&Q roughly means "do P then Q".

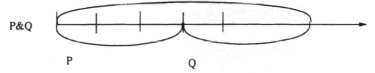

Using the chop operator we can express sometime ◇ and always □.

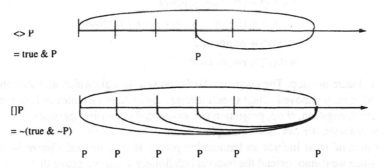

A projection operator creates coarse grain time using a repeated interval. *P proj Q* means Q is true on a coarse grain time interval. In this interval, clock ticks are defined by the repetition of *P*.

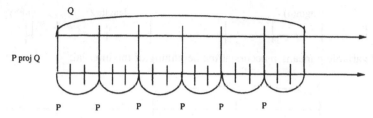

We shall use the following abbreviations,

$$P \vee Q \equiv (\neg P) \Rightarrow Q$$

$$P \wedge Q \equiv \neg(P \Rightarrow \neg Q)$$

$$P \Leftrightarrow Q \equiv (P \Rightarrow Q) \wedge (Q \Rightarrow P)$$

$$more \equiv \neg empty$$

$$+P \equiv P\&(P \vee empty)\&...\&(P \vee empty)$$

$$\Diamond P \equiv T\&P$$

$$\Box P \equiv \neg\Diamond\neg P$$

$$\bigcirc P \equiv @P \vee empty$$

$$skip \equiv @empty$$

$$length(n) \equiv \underbrace{@@...@}_{n} empty$$

$$less(n) \equiv \underbrace{\bigcirc\bigcirc...\bigcirc}_{n} F$$

$$\forall P \ f(P) \equiv \neg\exists P\neg f(P)$$

$$P\&\&Q \equiv (P \wedge \neg empty)\&Q$$

$$* P \equiv (P \ proj \ T) \vee (empty \wedge P) \ (closure)$$

$$fin(P) \equiv empty \Rightarrow P$$

$$halt(P) \equiv empty \Leftrightarrow P$$

$+P$ is a closure of chop. The *chop standard form* is a formula which all these abbreviations have been removed. Chop standard form may include variables and conjunction, disjunction, negation, chop, projection and existential quantifier operations.

For example, we can make a simple theorem, $\Diamond empty$, since we use finite interval (every interval must include an termination point). Hence, its dual $\Box more$ is unsatisfiable since we cannot extend the interval indefinitely. Later we prove that

$$(\Box\Diamond P) \Leftrightarrow (\Diamond\Box P) \Leftrightarrow fin(P),$$

from which we deduce ITL cannot express fairness. As indicated in [13], the decision procedure is simple for finite intervals.

1.1   Specification in Interval Temporal Logic

In ITL, it is easy to express sequential execution and time out. For example

$$((less(5) \wedge \Diamond p \wedge \Diamond q) \vee (length(6)\&s))\&\Box r$$

This means that p and q have to be done in 5 clock cycles, and after that r stays true until the end of the interval. Otherwise s is happen before r.

Using *proj*, the repeated event and time sharing task are easily described as in [3]. The expression

$$(length(2) \wedge \Diamond p) \; proj \; T$$

represents a process in which p happens every 2 clock cycles (its timing is not specified).

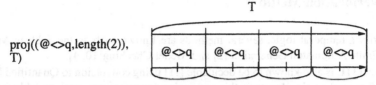

proj((@<>q,length(2)), T)

Conversely some preemptable task p which takes 10 ticks can be represented as follows

$$T \; proj \; (length(4) \wedge \Box p)$$

proj(T, ([]p,length(4)))

Of course, we can add a time limit easily. For example, if task p has to be done before q will happen:

$$((T \; proj \; (length(4) \wedge \Box p)) \wedge keep(\neg q)) \& q.$$

For a more complex example, if we have 2 periodical tasks (intervals are 3 clocks and 5 clocks) and one time sharing task with dead line with *length*(15), which shares one resource. (Fig.3.1).

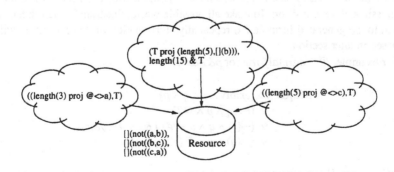

Fig. 3.1. Real-time Task Combination

$$((T \; proj \; (length(5) \wedge \Box(c))) \wedge length(15))\&T) \wedge$$
$$((((length(3) \; proj \quad @\Diamond(a)) \wedge T)\&less(3)) \wedge$$
$$((((length(5) \; proj \quad @\Diamond(b)) \wedge T)\&less(5)) \wedge$$
$$\Box(\neg((a \wedge b))) \wedge$$
$$\Box(\neg((b \wedge c))) \wedge$$
$$\Box(\neg((c \wedge a)))$$

2   Verification Methods

To verify a temporal logic, several methods are known such as the Tableau Method [14], Finite Automaton Generation [2] and Model Checking[10, 4].

Local ITL is also known to be decidable [11] using conversion to Quantified Linear Time Temporal Logic. However, this conversion generates one extra variable for each & operator which makes verification space/time hard. It also introduces infinite the interval and fairness unnecessarily. Hereafter we restrict ourselves in Local ITL. There is a model checker for Branching Temporal Logic / Computation Tree Logic [10] which has polynomial order complexity. However this is not a complete verifier.

3   Deterministic Tableau Expansion

In ITL, a temporal logic formula P can be separated into two parts: the current clock period and the future clock period. This separation can be represented by a disjunctive normal form with the *empty* and the @ (strong next) operators.

$$\vdash P \Leftrightarrow (empty \wedge P_e) \vee \bigvee_i P_i \wedge @Px_i$$

A formula P is true on an empty interval if P_e is true. In the case of a non-empty interval, the required condition Px_i at the next clock period depends on the current state condition P_i. P_e and P_i must not contain temporal logic operator. We call this separated form the $@-normalform$. Each P and Px_i represents a possible world, and which are connected by a possible clock transition. To make all possible world, this transformation has to be applied to the generated formula Px_i repeatedly. Termination of this procedure will be discussed in later section.

For example, $@-normalform$ for $p\&q\&r$ is

$$\vdash p\&q\&r \Leftrightarrow (empty \wedge r \wedge q \wedge p)$$
$$\vee (r \wedge q \wedge p \wedge @T)$$
$$\vee (\neg(r) \wedge q \wedge p \wedge @(T\&r \vee T\&q\&r))$$
$$\vee (\neg(q) \wedge p \wedge @(T\&q\&r)).$$

This $@-normalform$ represents a non-deterministic state transition shown in Fig.3.2.

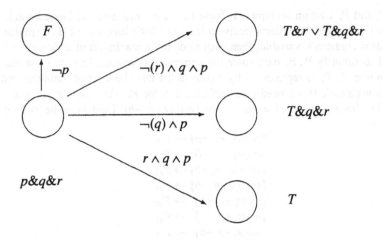

Fig. 3.2. State Transition for Chop Operator

This separation is performed recursively on temporal logic operators in the formula. For example, if we have two @-normal forms for P and Q then,

$$P = (empty \wedge P_e) \vee \bigvee_i P_i \wedge @Px_i$$

$$Q = (empty \wedge Q_e) \vee \bigvee_i Q_i \wedge @Qx_i$$

The @-normal form for $P\&Q$ will be,

$$P\&Q = (empty \wedge P_e\&Q) \vee \bigvee_i P_i \wedge @Px_i\&Q.$$

The expansion is easy because we use non-deterministic state transitions, but there is a problem. Since we use @-normal form (which is a kind of disjunctive normal form) negation becomes expensive. If P contains n disjunction then n-times normalization is necessary to achieve @-normal form. This corresponds the fact that this transformation generates non-deterministic transition.

However, if the conditions P_e, P_i do not overlap each other (i.e. if the transition conditions P_e, P_i are deterministic) negation becomes very easy,

$$\vdash \neg P \Leftrightarrow (empty \wedge \neg P_e) \vee \bigvee_i P_i \wedge @\neg Px_i. \text{ (if } P_e, P_i \text{ do not overlap each other)}$$

We call @-normal form deterministic if the conditions P_e and P_i do not overlap. Fortunately, it is possible to keep deterministic @-normal form in every tableau expansion of an ITL operator.

Since P_i and P_e contain no temporal logic formulae then non-overlapped conditions can be represented as a binary decision tree, in which leaves are ITL formulae. If the condition contains n variables then each node has a maximum of 2^n leaves. We do not need to simplify P_i, P_e part, since the expansion is unique. In fact, for a binary decision tree, P_i, P_e is represented by a path in the tree (i.e a set of variables and *empty* or its negation). If we need two variables a, b for P_e, the possible paths are: $[empty, +a, +b]$, $[empty, +a, -b]$, $[empty, -a, +b]$, $[empty, -a, -b]$. Then we write @-form for P like this:

$$P : \quad [empty, +a, +b] \rightarrow P_{e0}$$
$$[empty, +a, -b] \rightarrow P_{e1}$$
$$[empty, -a, +b] \rightarrow P_{e2}$$
$$[empty, -a, -b] \rightarrow P_{e3}$$
$$[more, +a, +b] \quad \rightarrow P_{x0}$$
$$[more, +a, -b] \quad \rightarrow P_{x1}$$
$$[more, -a, +b] \quad \rightarrow P_{x2}$$
$$[more, -a, -b] \quad \rightarrow P_{x3}$$

$\rightarrow$ means a state transition here. P_{ei} are T or F because it contains no temporal logic operator or variables. P_{xi} are temporal logic formulae, which label possible worlds as states. In this way, the tableau expansion can generate a deterministic automaton. To check the finiteness of the automaton, another normal form technique is necessary for the leaves (which will be discussed in a later section).

For fixed P_i, P_e, the deterministic tableau expansion rules can be described as a boolean operation on the leaves. Here we assume P's leaf for a P_i condition is $more(P)$ and P's leaf for a P_e condition is $empty(P)$. If we meet a local variable p, a node is added to the binary decision tree, that is, P_i is changed into two leaves $P_i \wedge p$ and $P_i \wedge \neg p$. Since $empty(P)$ contains no ITL operators, no variables and no connectives, $empty(P)$ is T or F.

T

$$empty(T) \quad = T$$
$$more(T) \quad\quad = @T$$

$P \wedge Q$

$$empty(P \wedge Q) = empty(P) \wedge empty(Q)$$
$$more(P \wedge Q) \;\; = more(P) \wedge more(Q)$$

$P \vee Q$

$$empty(P \vee Q) = empty(P) \vee empty(Q)$$
$$more(P \vee Q) \;\; = more(P) \vee more(Q)$$

$\neg P$

$$empty(\neg P) \quad = \neg empty(P)$$
$$more(\neg P) \quad\;\; = \neg more(P)$$

$@P$

$$empty(@P) \quad = F$$
$$more(@P) \quad\;\; = @P$$

$P \& Q$

$$empty(P \& Q) = empty(P) \wedge empty(Q)$$
$$more(P \& Q) \;\; = (empty(P) \wedge more(Q)) \vee (more(P) \& Q)$$

$\exists yQ$ y is removed from leaf conditions

$$empty(\exists yQ) = (empty(y \wedge Q) \vee empty(\neg y \wedge Q))$$
$$more(\exists yQ) = \exists y((more(y \wedge Q) \vee more(\neg y \wedge Q)))$$

$*(P)$

$$empty(*(P)) = empty(P)$$
$$more(*(P)) = more(P)\& * (P)$$

$P \ proj \ Q$

$$empty(P \ proj \ Q) = empty(Q)$$
$$more(P \ proj \ Q) = more(P)\&(P \ proj \ more(Q))$$

These transformation rules are part of the complete axiom system in ITL. To see the soundness of these transformations, we have to look at the model definition of the temporal logic operator.

For the chop rule, we have

$$M_{ij}(P\&Q) = T \text{ when } i \le \exists k \le j,$$
$$M_{ik}(P) = T, M_{kj}(Q) = T$$
$$F \text{ otherwise.}$$

First we note that $M_{ij}((P \vee Q)\&R) = T$ if $M_{ij}(P\&R) = T$ or $M_{ij}(Q\&R) = T$. Since leaves in a binary tree are all connected by disjunctions, a proof on a leaf is sufficient.

In the case of empty, $i = j$ in $M_{ij}(P\&Q) = T$, then i can be used as a k, and $M_{ii}(P) = T$ and $M_{ii}(Q) = T$. This is equivalent to the $M_{ii}(P) \wedge M_{ii}(Q)$. That is $empty(P\&Q) = empty(P) \wedge empty(Q)$. If $j > i$ then $k = i$ or $k > i$. In case of $k = i$, $M_{ik}(P) = T$ and $M_{kj}(Q) = T$ are necessary, so that $empty(P) = T$ and $more(Q)$. Otherwise $k > i$, requires $more(P)\&Q$. QED.

Since all the mapping function has distribution rule for disjunction, other rules can be proved in the same way. This is because ITL's chop operator has the same property as existential quantifier.

3.1 Expansion Example

The tableau expansion of $p\&q$ generates a tree with 6 leaves. For the empty condition we can replace & with $\wedge$. Then we have

$$empty \wedge (p\&q) : \quad [empty, +q, +p] \to T$$
$$[empty, -q, +p] \to F$$
$$[empty, -p] \quad \to F.$$

For the non-empty condition,

$$more \wedge (p\&q) : \quad [more, +q, +p] \to T$$
$$[more, -q, +p] \to (T\&q)$$
$$[more, -p] \quad \to F.$$

The first line comes from $empty(P \wedge Q) \vee (more(P)\&Q)$ and $empty(P \wedge Q) = T$. The second line comes from $empty(P \wedge Q) = F$ and $more(P) = T$.

3.2   Binary Subterm Tree

During possible world generation, various kind of ITL formulae are generated. Unlike LTTL or ETL [14], generated formulae contain more complex terms than the original subterm. It is not easy to see the finiteness of generated formulae.

To overcome this situation, we introduce a binary subterm tree. This subterm tree contains typed nodes:

- A triple $?(P, Q, R)$ is a binary decision node, in which if variable P is T then Q else R.
- $\exists x Q$, where x is a free variable.
- a numbered node for a unary temporal logic operator $O(P)$. (ex. @, *)
- a numbered node for a binary temporal logic operator $O(P, Q)$. (ex. &, *proj*)

Translation from ITL formula to binary subterm tree is done bottom-up. For example, $\diamond\square p$ is expanded into a chop standard form: $T\&\neg(T\&\neg p)$. First $\neg p$ is translated into,

$$?(p, F, T).$$

Then we need a numbered node s_1 for the chop operator, such that,

$$s_1 = T\&?(p, F, T).$$

Then the original formula is transformed into a numbered node, such that,

$$s_2 = T\&?(s_1, F, T).$$

After tableau expansion of this formula, we have a complex formula, $(\neg(T\&\neg p)) \vee (T\&\neg(T\&\neg p))$. But the result of the transformation is simple (Fig. 3.3),

$$s_3 = ?(s_2, T, ?(s_1, F, T)).$$

In this way, we can store generated formulae compactly in a binary subterm tree. As with the binary decision diagram, if we fix the ordering of nodes from top to bottom, the form of a node becomes unique to the logical connectives such as negation, conjunction or disjunction.

3.3   Termination of tableau expansion

If binary subterm trees contain finite numbered nodes, a set of the binary subterm trees must be finite, During the tableau expansion, we generate a formula which contains temporal logic operators. If this generated formula contains a new form of binary subterm tree in the argument of the operator then it may require a new node.

The expansion rules for logical connectives and the operator do not increase numbered nodes. Other rules generate only a fixed amount of temporal logic operators. Although we do allow recursion, we ensure that the depth of the temporal logic operator is monotonically decreased in an argument. For example, in projection the former part decreases the depth monotonically but the latter part increases by a single & operator.

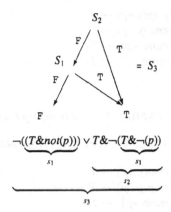

Fig. 3.3. Binary Subterm Tree

Subsequently, there are no infinitely applied rules, and only a finite number of new temporal logic operators are generated.

Here we prove this for the *proj* operator. Others can be proved in the same way.

Suppose *more(P)*, *more(Q)* generate finite variant. In the tableau expansion of *proj* ,

$$more(P \; proj \; Q) = more(P)\&(P \; proj \; more(Q)),$$

it generates a new node for the chop operator and the projection operator. The former part of the chop *more(P)* can vary according to the variant. The latter part of the chop can vary according to the variant of *more(Q)*. So the number of generated formulae is less than the number of products of variants for *more(P)* and *more(Q)*. QED.

Since the tableau expansion generates a finite number of binary subterm tree nodes, it generates only a finite binary subterm tree. When we expand all binary subterm trees, the expansion completes.

3.4 Verification Example

First try to prove $fin(p) \Leftrightarrow \Box(\Diamond p)$. The chop normal form is rather complex:

$$(\neg(T\&(empty \wedge p)) \vee \neg(T\&\neg(T\&p))) \wedge (T\&(empty \wedge p) \vee T\&\neg(T\&p))$$

We number this state as S_1. It looks complex but it has a simple binary subterm tree form. It has only three &, so we have three binary subterm nodes:

$$
\begin{aligned}
s_1 &: T\&?(p, ?(empty, T, F), F)) &&= T\&(empty \wedge p) \\
s_2 &: T\&?(p, T, F)) &&= T\&p \\
s_3 &: T\&?(s_2, F, T)) &&= T\&\neg(T\&p).
\end{aligned}
$$

The original formula has binary subterm tree: $?(s_3, ?(s_1, F, T), ?(s_1, T, F))$.

It is translated into @-normal form.

$$S_1 : \quad [empty, +p] \rightarrow T$$
$$[empty, -p] \rightarrow T$$
$$[more, +p] \rightarrow S_1$$
$$[more, -p] \rightarrow S_2$$

S_2 is a newly generated state,

$$(\neg(T\&(empty \land p)) \lor \neg((\neg(T\&p) \lor T\& \neg(T\&p)))) \land (T\&(empty \land p) \lor \neg(T\&p) \lor T\& \neg(T\&p)).$$

Its subterm tree form is $?(s_3, ?(s_1, F, T), ?(s_2, ?(s_1, T, F), ?(s_1, F, T)))$, and it is expanded into:

$$S_2 : \quad [empty, +p] \rightarrow T$$
$$[empty, -p] \rightarrow T$$
$$[more, +p] \rightarrow S_1$$
$$[more, -p] \rightarrow S_2$$

This is exactly the same as S_1's transition function. There are no newly generated formulae in this case, so we can finish the tableau expansion procedure. It is easy to check every empty leaf has T, so the original formula is valid.

4   Generated Model and Counter Example Generation

After tableau expansion, we have a deterministic finite automaton. Each state in an automaton is labelled by a unique binary subterm tree. The transition condition of the automaton is a list of +variable and -variable. The initial state is labelled by the original formula, and the empty transition (i.e. transition in which condition contains the empty operator) generates T or F.

This automaton returns T or F for finite series of events represented by local interval temporal logic variables. That is, it characterizes the original temporal logic formula. Since we are handling finite sequences, the termination condition of an automaton is a part of the assumption.

The output automaton state indicates which subterm is true or false; it can be used as a specification tester in hardware implementation.

If for all transitions which contain empty results T then it must accept all possible sequences of truth value assignments for the variables. That is, the original formula is valid. If there are no T nodes in the empty transition then the original formula is unsatisfiable.

Once we have an F node in the empty transition, it is easy to generate a counter example. The problem is how to discover a path from the initial state to the F node. The shortest path is easily found, if we reverse all links and mark the states in the automaton with a number in generated order.

First we check the least numbered F node, then pick up the least numbered node from the reverse links from the F node. We repeat tracing the least numbered node in the reverse links. Eventually we will reach the initial node which has the least numbered node in the root of reverse links. The generated shortest path should be acyclic. The traced path represents the shortest counter execution. If we start from a T node, we will have the shortest sample execution. Unfortunately, the shortest examples need not necessarily be useful.

4.1 Execution Examples

Consider next formula:

$$(p\&\&p\&\&p\&\&p\&\&p\&\&\neg p\&\&p) \rightarrow \Box(\Diamond p).$$

Using our verification program, it generates 13 states, 13 subterms, and 42 state transitions. It takes 13.7 sec on T2200SX, 386sx IBMPC compatible Laptop. There is a

```
| ?- diag.
counter example:
0:+p 2
1:+p 3
2:+p 5
3:+p 7
4:+p 9
5:-p 11
6:+p 12
7:-p F
```

Fig. 3.4. Example of Counter Example Generation

counter example because p is a local variable and there are no constraints on the end of an interval on the assumption in the formula. It is possible to make p false at the end of an interval which violates $\Box\Diamond p$, that is $fin(p)$.

The next one is more complex.

$$(((T\ proj\ (length(5) \wedge \Box(dc))) \wedge length(15))\&T)\wedge$$
$$((length(3)\ proj\ @\Diamond(ac)) \wedge T)\wedge$$
$$((length(5)\ proj\ @\Diamond(bc)) \wedge T)\wedge$$
$$((length(5)\ proj\ @\Diamond(cc)) \wedge T)\wedge$$
$$\Box(((ac \wedge \neg(bc) \wedge \neg(cc) \wedge \neg(dc))\vee$$
$$(\neg(ac) \wedge bc \wedge \neg(cc) \wedge \neg(dc))\vee$$
$$(\neg(ac) \wedge \neg(bc) \wedge cc \wedge \neg(dc))\vee$$
$$(\neg(ac) \wedge \neg(bc) \wedge \neg(cc) \wedge dc)\vee$$
$$(\neg(ac) \wedge \neg(bc) \wedge \neg(cc) \wedge \neg(dc))))$$

The first line is a time sharing task which has a deadline $length(15)$ and requires 5 clock cycles to be done. Between the second and the fourth lines are 3 periodical tasks. The reminder consists of shared resource conditions, that is, ac, bc, cc and dc cannot happen together. If it has no possible execution, these real-time tasks are not schedulable.

Using the verifier, it generates 242 states, 115 subterms and 799 state transitions. It takes 336.591 sec on 386sx 20Mhz, and it finds there are no possible executions. If

we change the load on time sharing task from 5 clock cycle to 4 clock cycle it finds
a possible execution. This time it requires 392.28 sec and it generates 303 states, 107
subterms and 982 state transitions. A generated possible execution is shown in Fig. 3.5.

```
execution:
0:-ac-bc-cc+dc 2
1:-ac-bc-cc+dc 3
2:-ac-bc-cc+dc 8
3:+ac-bc-cc-dc 13
4:-ac+bc-cc-dc 16
5:-ac-bc+cc-dc 19
6:+ac-bc-cc-dc 20
7:-ac-bc-cc+dc 21
8:+ac-bc-cc-dc 27
9:-ac+bc-cc-dc 31
10:-ac-bc+cc-dc 33
11:+ac-bc-cc-dc 35
12:-ac+bc-cc-dc 50
13:+ac-bc-cc-dc 52
14:-ac-bc+cc-dc 54
15:-ac-bc-cc+dc 0
```

Fig. 3.5. A Possible Execution

The implementation includes X window Interface as shown in Fig. 3.6.

5   Combination of Clausal Form Programs and Constraints

The first order version of the Interval Temporal Logic interpreter is called Tokio. The
relationship between Tokio and first order ITL is just like the one between Prolog and
first order predicate logic. If we do not use temporal operators, the execution of Tokio
is just the same as that of Prolog.

The syntax of Dec-10 Prolog is chosen here [1].

$$ap([], X, X). \tag{3.1}$$

$$ap([H||X], Y, [H||Z]) :- ap(X, Y, Z). \tag{3.2}$$

This is a Prolog append program. [X|Y] means a cons pair and [] means nil. A form
ap(...) is called a predicate. Variables in Tokio are words beginning with capital letters
or _.

A Tokio program: $p1, p2, ..., pn$ is true in the interval *int* is described as:

$$int : -p1, p2, ..., pn.$$

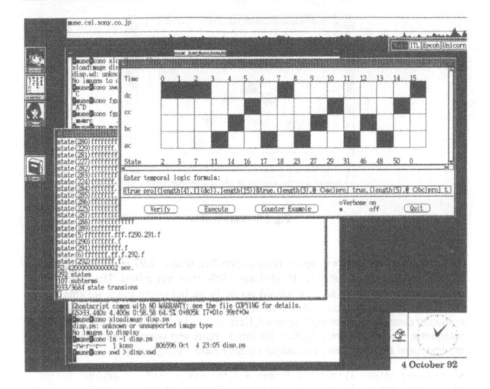

Fig. 3.6. X window display

That is, the interval name is used as a predicate name, and p1, p2, ..., pn are used as bodies. The Horn clause in Tokio is extended with temporal operators, such as

$$t1 : -\Box write(0), length(3)\&\&\Box write(1), length(5).$$

This writes four '0' and six '1' for each clock period.

6 Combination Example

Let's consider a toy GUI (Graphical User Interface) (Fig. 3.7). There are two buttons named start and stop and one signal. When start button is pushed the signal turns green and the GUI starts moving a small ball. When stop button is pushed the signal turns red and the ball stops. The specification is expressed in ITL like this:

$$+(((stop \land keep((red \land \neg(start))) \lor start \land keep((green \land \neg(stop))))))\land$$
$$\Box((red \land \neg(green) \lor \neg(red) \land green))\land$$
$$\Box((green \rightarrow move))\land$$
$$\Box((red \rightarrow \neg(move)))halt(quit)$$

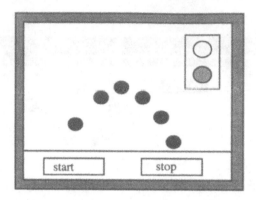

Fig. 3.7. A toy GUI

This is translated into state diagram using our verifier. It takes 2.97 sec in 386SX 20Mhz IBMPC and generates 3 states, 10 subterms, 13/90 state transitions. The specification contains many variables, so state transitions are large, but actual states are small.

The resulting state transitions are automatically translated into Tokio program. Generation of clausal form is straight forward. Here we use * for static variables in the Tokio interpreter. The events are classified for input events and output events. Input events are translated into equations and output events become assignments.

```
s1 :- empty,*stop= 1,*quit= 1,*green:= 0,*move:= 0,*red:= 1,empty.
s1 :- empty,*stop= 1,*quit= 1,*green:= 1,*move:= 1,*red:= 0,empty.
s1 :- empty,*stop= 0,*start= 1,*quit= 1,*green:= 0,*move:= 0,*red:= 1,empty.
s1 :- empty,*stop= 0,*start= 1,*quit= 1,*green:= 1,*move:= 1,*red:= 0,empty.
s1 :- more,*stop= 1,*start= 0,*quit= 0,*green:= 0,*move:= 0,*red:= 1,@s2.
s1 :- more,*stop= 0,*start= 1,*quit= 0,*green:= 1,*move:= 1,*red:= 0,@s3.
s2 :- empty,*quit= 1,*green:= 0,*move:= 0,*red:= 1,empty.
s2 :- empty,*quit= 1,*green:= 1,*move:= 1,*red:= 0,empty.
s2 :- more,*stop= 1,*start= 0,*quit= 0,*green:= 0,*move:= 0,*red:= 1,@s2.
s2 :- more,*stop= 0,*start= 0,*quit= 0,*green:= 0,*move:= 0,*red:= 1,@s2.
s2 :- more,*stop= 0,*start= 1,*quit= 0,*green:= 1,*move:= 1,*red:= 0,@s3.
s3 :- empty,*quit= 1,*green:= 0,*move:= 0,*red:= 1,empty.
s3 :- empty,*quit= 1,*green:= 1,*move:= 1,*red:= 0,empty.
s3 :- more,*stop= 0,*start= 1,*quit= 0,*green:= 1,*move:= 1,*red:= 0,@s3.
s3 :- more,*stop= 0,*start= 0,*quit= 0,*green:= 1,*move:= 1,*red:= 0,@s3.
s3 :- more,*stop= 1,*start= 0,*quit= 0,*green:= 0,*move:= 0,*red:= 1,@s2.
```

We can test the generated implementation for a set of input events using a Tokio program. Next program generates a start event and a stop event.

```
test :-
        static([green, red, move, stop, start]),
        *green:=0, *red:=1, *move:=0, *stop := 1, *start :=0 , *quit := 0
        && ((
            length(3), *stop := 0, *start :=1 &&
            length(3), *stop := 0, *start :=0 &&
```

```
              *stop := 1, *start :=0, @(*quit := 1)
           ),
           s1,
           [](((G= *green, R= *red, S= *start, P= *stop, M = *move,
                write((green,G,start,S,red,R,stop,P,move,M)))))).
```

Here is a possible execution result.

```
?- tokio test.

t0:
t1:green,0,start,0,red,1,stop,1,move,0
t2:green,0,start,1,red,1,stop,0,move,0
t3:green,1,start,1,red,0,stop,0,move,1
t4:green,1,start,1,red,0,stop,0,move,1
t5:green,1,start,0,red,0,stop,0,move,1
t6:green,1,start,0,red,0,stop,0,move,1
t7:green,1,start,0,red,0,stop,0,move,1
yes
| ?-
```

The example looks small but it becomes large quickly if we add more time constraints. But our verifier is robust for large rules and small number of variables. Since GUI prefer small number of buttons to control complex control such as double click, it seems that this method fits for GUI application generator. Especially simple adding of temporal constraints are far easier than modifications of the state diagram, which is buggy part of GUI.

A part of actual GUI code is shown below. This GUI uses InterViews package of SICStus Prolog.

```
toy :-  static([green, red, move, stop, start,quit]),
        *move:=1,*quit:=0,*stop:=0,*start:=1,*red:=0,*green:=1,
        bounce_init(W,R,G),@toy1(W,R,G).
toy1(W,R,G) :-
           [](event),                                     % input
           s1,                                            % automaton
           [](((button_red(R),button_green(G),bounce(W)))). % output

event :- nextevent(E),E=E1,event_select(E1).
event_select(noevent) :- true.
event_select(button(_,start)) :- *start := 1, *stop  := 0.
event_select(button(_,stop)) :-  *stop  := 1, *start := 0.
event_select(button(_,quit)) :-  *quit := 1.

button_red(Out) :-
        *red =0, Out => out("").
button_red(Out) :-
        *red =1, Out => out("Red").
```

event/0 generates input variables such as *start, while s1/0 handles state transition. button_red/1 is an object which handles necessary display procedures. Unlike other GUI packages, event handling is separated from GUI objects such as button or bounce procedure. Since GUI is inherently parallel, a combination of temporal logic and object oriented graphic package seems useful.

7   Future Work

To make things faster, since the subterm vector is used in the BDD implementation, it is possible to use a set of BDDs rather than a single huge BDD. A modification of existing state diagram is important, because synthesized state transitions can be very large.

References

1. W. Clocksin and C. Mellish. *Programming in Prolog*. Springer-Verlag, 1981.
2. G. de Jong. An Automata Theoretic Approach to Temporal Logic. In *Computer Aided Verification*. Springer-Verlag, July 1991. 3rd International Workshop, CAV'91.
3. R. Hale. Temporal logic programming, 1988.
4. H. Hirahashi. Design Verification of Sequential Machines Based on a Model Checking Algorithm of ε-free Regular Temporal Logic. Technical Report CMU-CS-88-195, Department of Computer Science, Carnegie-Mellon University, September 1988.
5. S. Kono, T. Aoyagi, M. Fujita, and H. Tanaka. Verification of Temporal Logic Programming Language Tokio. In *Logic Programming Conference '86*, 1986. (in Japanese).
6. S. Kono. Automatic Verification of Interval Temporal Logic. Technical Report SCSL-TM-92-007, Sony Computer Science Laboratory Inc., October 1992.
7. S. Kono. Automatic verification of interval temporal logic. In *8th British Colloquium For Theoretical Computer Science*, March 1992.
8. G. Milne. CIRCAL: A Calculus for Circuit Description. *Integration*, Vol. 1, No. 2, pp. 121–160, 1983.
9. R. Milner. *A Calculus of Communicating Systems*, volume 92 of *Lecture Note in Computer Science*. Springer-Verlag, 1980.
10. B. Mishra and E. Clarke. Automatic and hierarchical verification of asynchronous circuits using temporal logic. Technical Report CMU-CS-83-155, Dept. of Computer Science, Carnegie-Mellon Univ., September 1983.
11. B. Moszkowski. Reasoning about digital circuit. Technical Report No.STAN-CS-83-970, Dept. of C.S. Stanford Univ, July 1983.
12. N. Rescher and A. Urquhart. *Temporal Logic*. Springer-Verlag, 1971.
13. R. Rosner and A. Pnueli. A choppy logic, 1986.
14. P. Wolper. Synthesis of communicating processes from temporal logic specifications. Technical Report STAN-CS-82-925, Stanford University, 1982.

Temporal Logic and Annotated Constraint Logic Programming

Thom Frühwirth

Abstract. We introduce a family of logic programming languages for representing and reasoning about time. The family is conceptually simple while covering substantial parts of temporal logic. Given a logic in our framework, there is a systematic way to make it executable as a constraint logic program. Thus we can study and compare various temporal logics and their executable fragments.

Our approach allows for different models of time, different temporal operators, and temporal variables for both time points and time periods. Formulas can be labeled with temporal information using annotations. In this way we avoid the proliferation of variables and quantifiers as encountered in first order approaches. Unlike temporal logic, both qualitative and quantitative (metric) temporal reasoning with time points (instants) and periods (temporal intervals) are supported. A Horn clause fragment of our temporal logic can be seen as annotated constraint logic programming language. This class of languages can be implemented by translation into a standard constraint programming language. Thus we can make our temporal logic executable.

This paper is a companion paper to [Fru94c], where an interpreter for annotated languages and their underlying logic is described.

1   Introduction

We want to derive a temporal programming language from a temporal logic. There are two approaches to temporal logic: Modal logic and first order logics. Modal logic approaches capture naturally the relative position of formulae with respect to an implicit current time by talking about past, present and future. For example,

$age(john, 18) \rightarrow G\ vote(john)$

means that if John is 18 now, he can vote in the future (from now on).

On the other hand, first order logic approaches naturally support absolute positions of formulae along the time line by making time explicit. They suffer, however, from a proliferation of temporal variables and quantifiers. E.g.,

$age(john, 18) \rightarrow \forall t(later(t, now) \rightarrow vote(john))$

Usually, the logic will be reified, i.e. there are predicates that relate object formulas (that are terms in the logic) to temporal entities. For example,

$HOLDS(born(john), 1962).$

In an "unreified" logic, formulas have their usual status, the temporal information is included by adding extra arguments to the predicates and introducing additional predicates. E.g.,

$born(john, 1962).$

We propose a logic that lies in-between the two approaches, while it keeps most of the advantages of both. We make time explicit but avoid the proliferation of temporal variables and quantifiers of the first order approach. The basic idea is straightforward and not really new: We start from first order logic (FOL) and add time by "labeling" formulas with temporal information. The pieces of temporal information are given by *temporal annotations* which say at what time(s) the formula to which they are applied is valid. The resulting temporal annotated logic can be regarded as a modal logic, where the annotations are seen as parameterized modal operators. Likewise, it can be seen as reified first order logic where temporal annotations correspond to binary relations between predicates and temporal information.

Conceptually, this approach is simple: Unlike the unreified approach the annotations account for the special status of time and if we ignore the annotations, we are left with ordinary FOL.

The underlying idea of making temporal logics executable is to separate the temporal from the nontemporal aspects. To the temporal part, a special deduction method is applied, while the nontemporal part corresponds to first order logic where standard deduction suffices. Usually, only a fragment of the modal logic can be implemented in such a way. In the case of annotated constraint logic, we can implement its Horn clause fragment as a constraint logic programming language.

Our approach does not make any presuppositions on the ontology of time or denotation of formulas. We can specify the ontology of choice by adding the appropriate constraints to the logic. As a consequence, time may be linear (one future), branching (many possible futures) or circular (infinitely branching), discrete or continuous (dense), bounded or unbounded on either end (finitely or infinitely stretching into past or future). As needed, atomic formulas may represent events, states, properties, processes, actions and so on.

For a specific instance of our logical framework, one will try to map the constraints into already implemented ones, so that no new solver needs to be written. This is one of the reasons why our temporal programming language can be easily implemented.

We will see that the Horn clause fragment of our logic is related to generalized annotated logic programs (GAPs) [KiSu92]. However the flavour is different: While in [KiSu92] annotations are truth values, we consider annotations as modal operators.

In the next section we give a basic introduction into temporal logic. In the section after, we propose a temporal logic where formulas are annotated with temporal information. Then we introduce annotated constraint logic programs as a generalization of GAPs and in the section after show how to implement our temporal language as an instance of this language scheme.

2 Temporal Logic

The starting point for our investigation is the temporal logic resulting from introducing two new connectives, 'since' (S) and 'until' (U), to first order logic [Gal87]. These two temporal binary operators were introduced by Kamp in 1968. Time is strict linear. Furthermore, predicate symbols are made flexible (their interpretation depends on time), while function symbols are rigid (time independent) as before. Kamp showed that any first order temporal formula (over strict linear time) is expressible in such a temporal logic.

The definition of S and U in first order logic is:

$$\text{since: } A \; S \; B \; at \; t \; \Leftrightarrow \; \exists r \, (r < t \wedge A \; at \; r \wedge \forall s \, (r < s < t \rightarrow B \; at \; s))$$
$$\text{until: } A \; U \; B \; at \; t \Leftrightarrow \exists r \, (t < r \wedge A \; at \; r \wedge \forall s \, (t < s < r \rightarrow B \; at \; s))$$

where A and B are temporal formulas and r, s and t are temporal variables denoting time points. The relation $<$ is an ordering relation on time points, i.e. $r < t$ means that r is strictly earlier than t. $A \; at \; t$ means that the formula A is true at time point t. We may think of a time point as denoting an indivisible, duration-less instant or moment of time. We call $at \; t$ a *temporal annotation* of the formula A. This temporal annotation distributes over the logical connectives.

Let lb (ub) be the lower bound (upper bound) of the temporal time line or $-\infty$ (∞) in case of an unbounded temporal model. We require that every temporal variable, say t, is constrained to be within these bounds, $lb \leq t \leq ub$. Typically, $lb = 0$ and $ub = \infty$.

The following commonly used modal operators (introduced in the Tense Logic of Prior in 1955) can be defined in terms of the basic connectives S and U as well as negation and the basic propositions 'true' and 'false'. Their first order definitions are:

sometime past:	$P \, A \; at \; t \Leftrightarrow \exists r \, (r < t) \wedge A \; at \; r$
always past:	$H \, A \; at \; t \Leftrightarrow \forall r \, (r < t \rightarrow A \; at \; r)$
sometime future (eventually):	$F \, A \; at \; t \; \Leftrightarrow \exists r \, (t > r \wedge A \; at \; r)$
always future:	$G \, A \; at \; t \Leftrightarrow \forall r \, (t > r \rightarrow A \; at \; r)$

If time is discrete, we also allow for:

previous:	$\bullet A \; at \; t \Leftrightarrow \forall r \, (p(t) = r \rightarrow A \; at \; r)$
next:	$\circ A \; at \; t \Leftrightarrow \forall r \, (s(t) = r \rightarrow A \; at \; r)$

where s (p) is the successor (predecessor) function.

In some approaches, the above temporal operators are defined using a non-strict ordering relation instead of a strict one. For example, Templog, a "temporal Prolog", [AbMa89] uses the unary temporal operators concerned with future from above, but the future includes the present, i.e.

sometime future: $F\,A\ at\ t \Leftrightarrow \exists r\,(t \geq r \wedge A\ at\ r)$

Some approaches consider only a subset of the above operators or do not distinguish between past and future in their operators. Our methodology for implementing temporal constraint logic programming languages is not affected by these and similar variants.

3 A Temporal Annotated Logic

Different from the first order approach, we will use set annotations to capture quantified temporal variables. The idea is to see that quantification over a temporal variable intentionally defines a (possibly infinite) set of time points. The set approach is not new, e.g. [McD82] uses a similar construction. For example, given (a current time) t, the expression $\forall r\,(r < t)$ refers to a set I of all time points r such that $r < t$. An existential quantification such as $\exists r\,(r < t)$ refers to a time point r that is an element of the same set I.

Note that the above sets of time points correspond to *time periods*[1]. In practice, time periods are often represented by intervals (e.g. to avoid storage and access time problems [GaMcB91]). Intervals are defined by their end-points, say r and s. The interval then includes all points between r and s. Intervals may be closed or open on either side. In the unbounded time model, we also allow ∞ and $-\infty$ as end-points. As intervals represent (convex) sets, we may use, from set theory, relations (like equality and inclusion) and operations (like union, intersection and complement) on intervals. These relations and operations can be efficiently implemented by comparison of and computation on the end-points of the intervals. However, intervals are not closed under union and complement. This problem is avoided by taking sets of intervals instead of intervals as the basic temporal data-structure.

Sets of intervals provide a compact, finite representation of finite and infinite sets of time points. For notational convenience, we write $[r, s]$ instead of $\{[r, s]\}$ and t instead of $[t, t]$[2]. The interval (set) $[r, s]$ is empty if $r > s$ and is uniquely represented by []. In the following, we will not distinguish between sets of time periods and the sets of time points they represent. For simplicity and without loss of generality, we will talk about single time periods instead of sets of time periods.

Let A, B be first order formulas, t be a time point and I, J, K be time periods. We relate a formula to a time period by introducing two additional temporal annotations, *in* and *th*. If a formula A holds at *some* time point(s) (but we don't know exactly when) in a time period I we write $A\ in\ I$. The first order definition of the annotated formula is:

$A\ in\ I \Leftrightarrow \exists t\,(t \in I \wedge A\ at\ t)$

If a formula A holds *throughout* a time period, i.e. at *every* time point in a time period I, we write $A\ th\ I$. The definition is:

$A\ th\ I \Leftrightarrow \forall t\,(t \in I \rightarrow A\ at\ t)$

---

[1] We use *(time) period* instead of *(temporal) interval* throughout, to avoid confusion with purely mathematical notion of interval.

[2] Another choice would be to dissallow intervals representing singleton sets.

Through this definitions, we relate time periods (*in* and *th*) to time points (*at*).

These three temporal annotations correspond to the temporal predicates proposed in [Gal90]. In particular, HOLDS-IN(A, I) is the same as A *in* I, HOLDS-ON(A, I) is the same as A *th* I, and HOLDS-AT(A, t) is the same as A *at* t. This is surprising, as Galtons motivation is to represent continuous change, while our motivation is to arrive at an implementable temporal logic that avoids explicit quantification.

The axioms of our temporal logic can be formally justified as consequences of the corresponding expressions in first order logic. Not all theorems here are theorems of [Gal90], as he does not assume the existence of singleton intervals. Contrary to his statement that "instants [time points] are not related to intervals [time periods] ... as members to sets" (p. 184), his axioms for WITHIN do not exclude this interpretation either. For example, if in Galtons examination

$$\forall r \exists I (\text{WITHIN}(r, I) \Rightarrow \forall s (\text{WITHIN}(s, I) \Rightarrow r = s))$$

is added, his axioms (SM) and (SP) would turn into theorems and thus states of motion and position not be distinguished by the theory.

First of all, if a time period denotes a single time point, the three temporal annotation coincide[3]:

    (1*a*) A *at* $t \Leftrightarrow A$ *in* $[t, t]$

and

    (1*b*) A *at* $t \Leftrightarrow A$ *th* $[t, t]$

For all t, I, J such that $t \in I \wedge t \in J$, we have that:

    (2*a*) A *th* $I \Rightarrow A$ *at* t

and

    (2*b*) A *at* $t \Rightarrow A$ *in* J

Furthermore,

    (3*a*) A *th* $[] \Leftrightarrow true$

and

    (3*b*) A *in* $[] \Leftrightarrow false$.

The annotations *in* and *th* are dual with regard to classical negation:

    (4*a*) $\neg(A$ *in* $I) \Leftrightarrow (\neg A$ *th* $I)$

and

    (4*b*) $\neg(A$ *th* $I) \Leftrightarrow (\neg A$ *in* $I)$

---

[3] If singleton intervals were excluded, the temporal annotations would not coincide.

This classical negation has been termed *ontological* negation in [KiSu89], the *epistemic* negation mentioned in this work can be defined as:

(5a)   *not* (A *in* I) ⇔ A *in* (−I)

and

(5b)   *not* (A *th* I) ⇔ A *th* (−I)

where − is the complement operator.

We can also show that if A holds throughout a period, it also holds throughout all sub-periods.

(6a)   A *th* I ⇔ ∀J (J ⊆ I → A *th* J)

Analogously, if A holds at some time in a period, it also holds at some time in all super-periods:

(6b)   A *in* I ⇔ ∀J (J ⊇ I → A *in* J)

Other theorems for *th* include (for *in*, the theorems are dual):

(7)   (¬A) *th* I ⇒ ¬(A *th* I)
(8)   (A ∧ B) *th* I ⇔ (A *th* I ∧ B *th* I)

We now discuss the relationship of this temporal logic with other work. We already have mentioned [Gal90]. This work was a critical examination of [All84]. Allen [All84] only considers time periods and no time points. His axioms imply dense, linear time and no singleton intervals (so that each time period has always a non-empty sub-period). His predicate HOLDS(P, I) (where P is a formula denoting a property) is equivalent to the annotated formula P *th* I. Hence all his axioms and theorems about HOLDS correspond to theorems derived from the first order definitions of the *th* annotation. As advocated in [Gal90], and different from [All84], we use classical negation. We have no correspondence to Allens OCCUR predicate for events (which has been shown to prevent certain intuitive conclusions in [Gal87]). Allens predicate OCCURING(P, I) (where P is a formula denoting a process) is similar to the annotated formula P *in* I. It is not equivalent, as the above-mentioned restriction on time periods means that there is more than one time point where P holds.

We can define the temporal operators of tense logic in terms of our temporal annotated logic. In this way, we avoid temporal variables and explicit quantification.

since:	A S B *at* t	⇔ ∃r (A *in* [r, t) ∧ A *at* r ∧ B *th* (r, t))
until:	A U B *at* t	⇔ ∃r (A *in* (t, r] ∧ A *at* r ∧ B *th* (t, r))
sometime past:	P A *at* t	⇔ A *in* [lb, t)
always past:	H A *at* t	⇔ A *th* [lb, t)
sometime future:	F A *at* t	⇔ A *in* (t, ub]
always future:	G A *at* t	⇔ A *th* (t, ub]
previous:	•A *at* t	⇔ A *at* p(t)
next:	∘A *at* t	⇔ A *at* s(t)

In temporal logic, a modal operator can be nested, as in $G\ F\ A\ at\ t$. Nested temporal operators cause temporal operators to be applied with reference to time periods (*in* and *th* annotations) as well. As above with *at* annotations, the right-hand side can sometimes be expressed as a conjunction of temporal annotated formulas, e.g. $\circ A\ th\ [r,s] \Leftrightarrow A\ th\ [s(r),s(s)]$. Nested temporal operators are a topic of our current research.

4 Annotated Constraint Logic Programs

Annotated constraint logic programming (ACLP) [Fru94c] combines generalized annotated logic programs (GAP) [KiSu89] with constraint logic programming (CLP) [VH91, F*92, JaMa94]. GAP and CLP programs can be mapped into each other, but with complications. The combination of the two approaches in ACLP has the advantage of conceptual simplicity. From GAP we get the annotations and from CLP the constraints. One advantage of ACL programs is that they can be implemented simply by a translation into existing constraint logic programming languages.

Like CLP, ACLP programs consist of a finite set of clauses and clauses are built from atoms and logical connectives. Atoms are a structure of the form $p(t_1,...t_n)$ where p is a predicate symbol of arity n ($n \geq 0$) and $t_1,...t_n$ is a n-tuple of terms. A term is a variable, e.g. X, or a structure of the form $f(t_1,...t_m)$ where f is a function symbol of arity m ($m \geq 0$) applied to a n-tuple of terms. Function symbols of arity 0 are also called constants, predicate symbols of arity 0 propositions. Predicate and function symbols start with lowercase letters while variables start with uppercase letters. In CLP, there is a distinguished class of predicates called *(relational) constraints*, and a distinguished class of interpreted functions called *functional constraints*. A *constraint theory* is the set of all sentences involving only relational and functional constraints (and no program-defined predicates). Equality (=) as well as *true* and *false* are relational constraints.

In GAP, literals can be annotated by a distinguished class of terms called *annotations*. Following [KiSu92], we assume an upper semi-lattice with ordering $\sqsubseteq$ (a transitive, reflexive and antisymmetric relation) and least upper bound operator $\sqcup$ (which is idempotent, associative and commutative). The semilattice needs not be complete. In ACLP, the annotations are terms. We write the annotation (term) immediately after the formula it annotates. The partial ordering $\sqsubseteq$ is a relational constraint. The least upper bound operator $\sqcup$ is a functional constraint

For example, the lattice may be the set of the real numbers R with the usual ordering $\leq$, upper bound operator is the maximum function, maximal element 1, minimal element 0. The functional constraints are $min, max, +, -$, and relational constraints are $=, \leq, <$. Infix notation may be used for relational constraints (e.g. $X \leq Y$) and functional constraints (e.g. $-X + Y$).

We can define an ACLP clause as follows:

$$H^a \leftarrow C_1 \wedge ... \wedge C_n \wedge B_1^{a_1} \wedge ... \wedge B_m^{a_m} \qquad (0 \leq n, m)$$

where H is the head atom, the B_i's are the body literals, the C_j's are the (relational) constraints, and a, a_k's are annotations[4]. ACLP clauses are similar to the constrained GAP clauses of [KiSu92]. If there are no annotations, the clause is a CLP clause.

We define the meaning of ACLP programs by translation into CLP programs. Each annotated atom $p(t_1, \ldots, t_n)^a$ is mapped into an atom $p(t_1, \ldots, t_n, a)$, using a translation function "*map*". Note the correspondence to unreifying a reified logic.

The above ACLP clause becomes:

$$map\ H^{A0} \leftarrow A0 \sqsubseteq a \wedge C_1 \wedge \ldots \wedge C_n \wedge map\ B_1^{a_1} \wedge \ldots \wedge map\ B_m^{a_m}.$$

This clause says that if H holds with some annotation a, then it also holds with all annotations $A0$ which are smaller in the lattice.

Furthermore, for each predicate p the following clause is added to guarantee the *closure* of the computed annotations:

$$p(X_1, \ldots, X_n, A1 \sqcup A2) \leftarrow p(X_1, \ldots, X_n, A1) \wedge p(X_1, \ldots, X_n, A2).$$

The above clause says that if a predicate holds with some annotation and the same predicate holds with another annotation, then the predicate also holds with the least upper bound of the annotations. This upward closure of annotations means that there is usually a single annotation that represents all the annotations for which a predicate holds. In general, the above clause will cause non termination. However, for many lattices, the clause can be specialised so that it terminates and computes the closure efficiently. The closure clause can even be omitted sometimes (e.g. for the lattice of natural numbers).

In [KiSu92], additional clauses are derived from existing clauses, the so-called re-ductants, to implement the GAP language. While reductants achieve the same as our closure clause, they result in a combinatorial explosion of the number of clauses in the program. Recently, "ca-resolution" for annotated logic programs was proposed in [LeLu94] and implemented in C. The idea is to compute dynamically and incrementally the reduction (that resulted in the reductants in [KiSu92]) by collecting partial answers. It turns out that operationally this is similar to our approach (which relies on recursion to collect the partial answers). However, in [LeLu94] the class of programs considered is smaller and the intermediate stages of a reduction are not sound with respect to the standard CLP semantics.

Continuing the example above, the following ACLP clauses with numeric annotations representing degree of belief

$rain : 0.9 \leftarrow grass\_wet : 0.8$

$rain : B \leftarrow clouds : B.$

are translated into CLP clauses

$rain(0.9) \leftarrow grass\_wet(0.8).$

$rain(B) \leftarrow clouds(B).$

$rain(max(A1, A2)) \leftarrow rain(A1) \wedge rain(A2).$

The last clause for closure can be omitted since $max(A1, A2)$ is either $A1$ or $A2$.

---

[4] In [KiSu89], there are some restrictions on the use of annotations in clauses, which we do not enforce in ACLP.

5 Temporal ACLP

In this section we define a Horn logic fragment of temporal annotated logic as an instance of ACLP. Thus a subset of temporal logic can be executed efficiently utilising state-of-the-art constraint logic programming technology. This temporal language has been implemented [Fru94c].

The implications in the theorems (2),(3) and (6) induce equivalence classes, a partial order and a least upper bound operator on the lattice of temporal annotations. For example, th $\{[1, 3], [2, 5]\}$ is the same as th $\{[1, 5]\}$ and greater than th $\{[2, 3]\}$, while in $\{[1, 3], [2, 5]\}$ is smaller than in $\{[2, 3]\}$ and greater than in $\{[1, 5]\}$. The least upper bound of annotations of the same kind is simply the union of their time periods.

Consequently, the maximal element in the lattice is the annotation in [], then follows the element th $[lb, ub]$, then follow all th annotations. In the "middle" of the lattice we have th $\{t\}$ = at t = in $\{t\}$ for all time points t. Then follow all in annotations. The minimal element in the lattice is the annotation th [], preceded by the element in $[lb, ub]$.

In our programming language, we require that annotations are applied to literals only. Without loss of generality we restrict annotations to single time periods for simplicity. An annotation A th $\{I_1, \ldots, I_n\}$ can be represented by A th $I_1 \wedge \ldots \wedge A$ th I_n, analogously for the in annotation.

Temporal ACLP clauses are translated into CLP clauses as in the general case. According to the topology of time needed, we may choose finite or infinite, bounded or unbounded sets of natural or real numbers to represent time periods. We only have to implement the ordering relation of the temporal lattice. We map it to inequalities between numbers. The ordering relation on the numbers is $\leq$, the relation $>$ results from negated in $\leq$. All variables refer to time points, except A, which stands for any annotation[5]:

$$th\ [S_1, S_2] \sqsubseteq A \qquad\qquad \leftarrow S_1 > S_2. \quad \% \text{ empty interval}$$
$$th\ [S_1, S_2] \sqsubseteq th\ [R_1, R_2] \leftarrow R_1 \leq S_1 \wedge S_2 \leq R_2.$$
$$th\ [S_1, S_2] \sqsubseteq at\ T \qquad\quad \leftarrow T = S_1 \wedge S_2 = T.$$
$$th\ [S_1, S_2] \sqsubseteq in\ [R_1, R_2] \leftarrow R_1 = R_2 \wedge R_1 = S_1 \wedge S_2 = R_2.$$
$$at\ T \qquad\ \sqsubseteq th\ [R_1, R_2] \leftarrow R_1 \leq T \wedge T \leq R_2.$$
$$at\ T_1 \qquad\ \sqsubseteq at\ T_2 \qquad\ \leftarrow T_1 = T_2.$$
$$at\ T \qquad\ \sqsubseteq in\ [R_1, R_2] \leftarrow R_1 = R_2 \wedge R_1 = T.$$
$$in\ [S_1, S_2] \sqsubseteq th\ [R_1, R_2] \leftarrow R_1 \leq T \wedge T \leq R_2 \wedge S_1 \leq T \wedge T \leq S_2.$$
$$in\ [S_1, S_2] \sqsubseteq at\ T \qquad\quad \leftarrow S_1 \leq T \wedge T \leq S_2.$$
$$in\ [S_1, S_2] \sqsubseteq in\ [R_1, R_2] \leftarrow R_1 \leq S_1 \wedge S_2 \leq R_2.$$

In addition, the closure clause can be specialized to produce *maximal* time periods for th annotations according to the equivalence relation:
$$p(X_1, \ldots, X_n, th\ [R, T]) \leftarrow R < S \wedge S < T \wedge$$
$$p(X_1, \ldots, X_n, th\ [R, S]) \wedge$$
$$p(X_1, \ldots, X_n, th\ [S, T]).$$

---

[5] The actual clauses are somewhat more contrived to guarantee termination and improve efficiency.

We now discuss some related work in temporal programming languages. In the companion paper [Fru94c], an interpreter-based implementation is given. Our work was motivated by [KiSu92], where an interval based logic is translated into GAPs. The annotations correspond to our *th* annotations. It is shown that the temporal logic of Shoham can be encoded as temporal GAP.

One of the first temporal logic programming languages was TEMPLOG, a "temporal Prolog" [AbMa89]. TEMPLOG implements a fragment of first-order temporal logic (tense logic). TEMPLOG is implemented using a special "temporal SLD resolution" strategy. This corresponds to a "direct" implementation approach which has the disadvantage that we have to start almost from scratch.

In [Brz93], a powerful temporal logic (tense logic extended by parameterized temporal operators) is translated into first order constraint logic. The resulting constraint theory is rather complex as it involves quantified variables and implication, whose treatment goes beyond standard CLP implementations. In Brzoska's programming language, temporal operators can be nested, but "eventuality" in the heads of clauses is dissallowed.

6 Conclusions

We defined a temporal annotated logic allowing for various models of time and various temporal operators for both time points and time periods (temporal intervals). Temporal annotated formulas avoid the proliferation of temporal variables and quantifiers of the first order approach while making temporal information explicit. A Horn clause fragment of the logic can be implemented as an annotated constraint logic programming language. In temporal ACLP, we can reason about qualitative and quantitative (metric), precise and fuzzy information about the absolute and relative location of literals annotated with time points and time periods along the time line. Temporal ACLP also contributes to the discussion of the relationship of temporal reasoning and GAP's [KiSu92].

This work can only be the starting point for implementing temporal logics by mapping them into our annotated temporal logic. This mapping and its limitations, as well as the extension of temporal ACLP to nested annotations and non-atomic literals, are currently investigated [Fru94b]. We are also considering extending the available temporal constraints along the lines of [Fru94a]. Another line of future work is to provide time periods as primitives, instead of deriving them from time points.

References

[AbMa89] M. Abadi and Z. Manna, Temporal Logic Programming, J. Symbolic Computation (1989) 8, pp 277-295.

[All84] J. F. Allen, Towards a General Theory of Action and Time, Artificial Intelligence, Vol. 23, 1984, pp 123-154.

[Brz93] C. Brzoska, Temporal Logic Programming with Bounded Universal Goals, 10th ICLP, Budapest, Hungary, MIT Press, 1993.

[F*92] T. Frühwirth et al., Constraint Logic Programming - An Informal Intro-duction, Chapter in Logic Programming in Action, Springer LNCS 636, September 1992. Also available by anonymous ftp from ftp.ecrc.de, in pub/ECRC_tech_reports/reports, file ECRC-9305.ps.Z.

[Fru94a] T. Frühwirth, Temporal Reasoning with Constraint Handling Rules, Technical Report ECRC-9405, ECRC Munich, Germany, January 1994. Available by anonymous ftp from ftp.ecrc.de, in pub/ECRC_tech_reports/reports, file ECRC-9405.ps.Z.

[Fru94b] T. Frühwirth, Annotating Formulas with Temporal Information, Workshop on Logic and Change at ECAI 94, Amsterdam, The Netherlands, August 1994.

[Fru94c] T. Frühwirth, Annotated Constraint Logic Programming Applied to Temporal Reas-oning, Programming Language Implementation and Logic Programming (PLILP), Madrid, Spain, Springer LNCS, September 1994. Also available by anonymous ftp from ftp.ecrc.de, in pub/ECRC_tech_reports/reports, file ECRC-94-22.ps.Z.

[Gal87] A. Galton (ed), Temporal Logics and Their Applications, Academic Press, 1987.

[Gal90] A. Galton, A Critical Examination of Allen's Theory of Action and Time, Artificial Intelligence, Vol. 42, 1990, pp. 159-188.

[GaMcB91] D. Gabbay and P. McBrien, Temporal Logic and Historical Databases, 17th Int. Conf. on Very Large Databases, pp 423-430, Barcelona, September 1991.

[JaMa94] J. Jaffar and M. J. Maher, Constraint Logic Programming: A Survey, Journal of Logic Programming, 1994:19,20:503-581.

[KiSu89] M. Kifer and V.S. Subrahmanian, On the Expressive Power of Annotated Logic Pro-grams, North American Conf. on Logic Programming, E.L. Lusk and R.A. Overbeek (eds), MIT Press, 1989, pp 1069-1089.

[KiSu92] M. Kifer and V.S. Subrahmanian, Theory of Generalized Annotated Logic Program-ming and its Applications, Journal of Logic Programming, April 1992.

[LeLu94] S. M. Leach and J. J. Lu, Computing Annotated Logic Programs: Theory and Im-plementation, 11th ICLP, Santa Margherita Ligure, Italy, MIT Press, 1994.

[McD82] D. McDermot, A Temporal Logic for Reasoning about Processes and Plans, Cognit-ive Science 6:101-155, 1982.

[VH91] P. van Hentenryck, Constraint Logic Programming, The Knowledge Engineering Review, Vol 6:3, 1991, pp 151-194.

Efficiently Executable Temporal Logic Programs

Stephan Merz

1 Introduction

Temporal logics (TL) of linear time [12, 15] offer very expressive languages that have been widely used for the specification and verification of reactive and concurrent systems. TL models describe not an unchanging world of logical truths, but an infinite sequence (of states) that naturally corresponds to the execution of a computer program over time. It is therefore a natural proposal to take a TL specification as described by a set of TL formulas, and "let it run". This requires an interpreter that accepts a set S of TL formulas as input, and generates a model of S that is then interpreted as a run of S. Different realizations of this idea can be found, among others, in [4, 11, 19, 16, 24]. Temporal logic programming languages have also been based on the proof-as-program paradigm, either using TL theorem provers [1, 6, 20]., or constraint logic programming languages [7, 9, 10].

The designer of a TL-based programming language is mainly faced with three interrelated problems:

Complexity Even for propositional temporal logics, the satisfiability problem, while decidable, is usually PSPACE-complete [22]. It is highly undecidable (Σ^1_1-complete) for first-order temporal logics [17, 23].

System vs. environment The interpreter is not allowed to change the state of the environment. This may rule out some models of the formula, and complicates the satisfiability problem for temporal logic programs.

Interactive runs A realistic model of computation for reactive systems requires incremental model construction where the system has to respond to a finite amount of environment input, and cannot backtrack on previous reactions that have already been observed by the environment.

It is therefore necessary to find some balance between restricting the fragment of TL that is acceptable as input to the interpreter, and the efficiency of the model construction algorithm that underlies the interpreter. In the following, I will try to shed some light on the issues underlying this tradeoff. In particular, in Sections 2 and 3, I will give a syntactic characterization of propositional safety properties in the fixed-point temporal logic νTL of [3, 5], and define a restricted class of safety properties that afford a linear model-construction algorithm. The choice of νTL is convenient from a theoretical point of view because of its simple syntax, and its high expressiveness. (In fact, all conventional TL connectives [12, 15] are definable in νTL.) For a practical TL-based programming language, one may want to restrict the actual syntax to some set of predefined temporal connectives with an intuitive semantics that does not involve explicit fixpoint constructions, or at least provide a set of such connectives as predefined macros. In Section 4, I will explain how environment conditions can be generated that guarantee successful program executions, using automata-theoretic techniques. In Section 5, I extend the formalism to the first-order case. Section 6 contains a comparison with related work.

2   Deterministic safety formulas

For simplicity, I will at first restrict attention to propositional TL. It is well known that every TL formula can be written as a conjunction of a safety and a liveness property [2, 14]. By definition, if L is a liveness property, then any finite state sequence may be extended to an infinite one that satisfies L. Consequently, liveness properties cannot guide an interpreter during the construction of single (or finitely many) states to extend an already produced prefix of some run. Insisting on incremental model construction, liveness requirements have to be implemented by explicit strategies (expressed as safety formulas) that, together with basic liveness properties of the programming model, and liveness assumptions on the environment, will ensure the desired behaviour.

To illustrate this point by a simple example, consider the specification of a transmitter module that sends data received from some other module along a channel. A specification like $\Box(DataReceived \rightarrow \Diamond TransmitData)$ presents the interpreter with a choice between sending the data immediately or delaying the transmission, which cannot be resolved locally. If the intended meaning of this specification is to transmit data as soon as possible, i.e. when the channel becomes free, this can—and should—be specified by a formula like $\Box(DataReceived \rightarrow TransmitData$ atnext $ChannelFree)$, which is a safety property.

It is therefore important to syntactically identify TL formulas that express safety properties. I will base the following discussion on the temporal logic νTL [3, 5], which is defined from proposition letters like $p, q, \ldots$ and proposition variables like $X, Y, \ldots$, using the next-time operator $\bigcirc$, and the greatest fixed point operator ν. For propositional νTL, the following fact has been shown in [18]:

Fact 1 *Let F be a formula of propositional νTL. F expresses a safety property iff there is a closed propositional νTL formula G such that G is equivalent to F, and all subformulas $(\nu X\, G)$ of F occur positively.*

Another, but related concern in the definition of an executable TL is the amount of nondeterminism that is present in the specification. Since the interpreter is required to operate without backtracking, any choices left to the interpreter have to be non-essential, in the sense that none of them may lead to failure to generate a model if another choice had led to success. But such choices may as well be resolved beforehand by the programmer. The following class of *deterministic safety formulas* attempts to formalize this intuition:

Definition 2
  - *The proposition constants* true *and* false *as well as the formulas p and $\neg p$, for any proposition letter p, are literals.*
  - *Every literal is an antecedent. If A and B are antecedents, then $A \wedge B$ and $A \vee B$ are also antecedents.*
  - *Every literal is a pre-DSF. If A is an antecedent, X is a proposition variable and F, G are pre-DSFs, then $F \wedge G$, $\bigcirc F$, $\bigcirc X$, $A \rightarrow F$ and $(\nu X\, F)$ are pre-DSFs.*
  - *Every closed pre-DSF is a deterministic safety formula (DSF).*

By fact 1, every DSF defines a safety property. It is interesting to note that DSFs characterize exactly the same class of properties as the *safety automata* of [14], which are deterministic and complete Streett automata that contain a designated failure state. However, to generate models of DSFs, the following alternative automata-theoretic representation of DSFs will be more useful:

Definition 3 *A DSF automaton $T = (N, I, S, \vartheta, \zeta, n_0)$ consists of:*

  - *a finite non-empty set N of nodes,*
  - *an acyclic relation $I \subseteq N \times N$ representing internal edges,*
  - *a relation $S \subseteq N \times N$ representing state transitions,*
  - *a node labelling ϑ assigning a finite set of literals $\vartheta(n)$ to every node $n \in N$,*
  - *an edge labelling ζ assigning an antecedent $\zeta(n, n')$ to every internal edge $(n, n') \in I$,*
  - *and a designated initial state $n_0 \in N$.*

A DSF automaton $T = (N, I, S, \vartheta, \zeta, n_0)$ is called

  - *normal if every edge label $\zeta(n, n')$ is a disjunction $L_1 \vee \ldots \vee L_k$ of literals,*
  - *regular if $I(n) = \varnothing$ or $S(n) = \varnothing$, for every node $n \in N$,*
  - *monotonic if $\vartheta(n) \subseteq \vartheta(n')$ whenever $(n, n') \in I$,*
  - *complete if for every node $n \in N$ such that $I(n) \neq \varnothing$, $\bigvee_{(n,n') \in I} \zeta(n, n')$ is (propositionally) valid.*

I will represent DSF automata graphically as in Figure 5.1, where internal edges are inscribed with the edge label, whereas "crossed" arrows are used for state transitions.

The following definition states what it means for a DSF automaton to accept a temporal structure $\sigma = (s_0, s_1, \ldots)$, where a state s_i is a Boolean valuation of proposition letters:

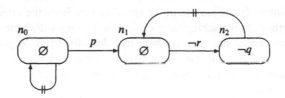

Fig. 5.1. Graphical representation of a DSF automaton

Definition 4 *The marking of a DSF automaton $T = (N, I, S, \vartheta, \zeta, n_0)$ for a temporal structure $\sigma = (s_0, s_1, \ldots)$ is the ω-sequence $M^\sigma = (m_0^\sigma, m_1^\sigma, \ldots)$ such that for every $i \geq 0$, $m_i^\sigma \subseteq N$ is the smallest set of nodes satisfying the following conditions:*

- $n_0 \in m_0^\sigma$
- *If $n \in m_i^\sigma$ and $(n, n') \in I$ is an internal edge such that $\sigma_i \models \zeta(n, n')$ then $n' \in m_i^\sigma$.*
- *If $n \in m_i^\sigma$ and $(n, n') \in S$ is a state transition then $n' \in m_{i+1}^\sigma$.*

σ is accepted by T if for every $i \in N$, the labels of all nodes $n \in m_i^\sigma$ hold at σ_i:

$$\sigma_i \models \bigwedge_{n \in m_i^\sigma} \vartheta(n) \quad \text{for every } i \in N.$$

For example, the DSF automaton of Figure 5.1 accepts exactly the models of the DSF

$$\Box(p \to \neg q \text{ unless } r) \equiv \nu X \, (p \to (\nu Y \, \neg r \to \neg q \wedge \bigcirc Y)) \wedge \bigcirc X$$

(By convention, the scope of ν in temporal formulas extends as far to the right as possible.)

In fact, DSFs and DSF automata agree in expressiveness:

Theorem 5
(i) *For every DSF φ there is a DSF automaton T^φ that accepts exactly the models of φ, and whose size is linear in the length of φ.*
(ii) *For every DSF automaton T there is a DSF φ^T whose models are exactly the temporal structures accepted by T.*

Proof: I only give the relevant constructions.

(i) The construction of a DSF automaton T^φ associated with a DSF φ is similar to the familiar tableau construction of temporal logic (see e.g. [26]). Formally, assume all proposition variables bound in different subformulas $(\nu X \, F)$ of φ to be pairwise different, and define the closure $cl(P)$ of any pre-clause P that appears as a subformula of φ as follows:

$cl(L) = \{L\}$ if L is a literal,

$cl(X) = \{X\} \cup cl(F_X)$ for a proposition variable X,

$cl(\bigcirc P) = \{\bigcirc P\}$, $cl(P \wedge Q) = \{P \wedge Q\} \cup cl(P) \cup cl(Q)$,

$cl(A \to P) = \{A \to P\}$, $cl(\nu X \, F_X) = \{(\nu X \, F_X)\} \cup cl(F_X)$, where F_X denotes the (unique) formula such that $(vX\, F_X)$ occurs as a subformula in φ. Start the construction of T^φ from an initial node n_0 labelled by $cl(\varphi)$ and inductively apply the following rules to construct new nodes and edges. Whenever a node with the required label already exists, only a new edge has to be inserted; if the two nodes are already connected by an edge, the new label will be added as a disjunct to the existing edge label:

(i) For every node n whose label contains a formula $A \to P$, construct a node n' labelled by $cl(P)$ and an internal edge $(n, n') \in I$ labelled by A.

(ii) For every node n whose label ϑ contains a formula $\bigcirc P$ construct a node n' labelled by

$$\bigcup\{cl(P) \mid \bigcirc P \in \vartheta\}$$

and a state transition $(n, n') \in S$.

T^φ is then obtained from the resulting graph by restricting all node labels to literals.

(ii) To construct φ^T for a DSF automaton $T = (N, I, S, \vartheta, \zeta, n_0)$ whose set of nodes is $N = \{n_0, \ldots, n_k\}$, let $X_0, \ldots, X_k$ be proposition variables, and define the formulas ψ_j^X and δ_j^X for $0 \le j \le k$ and every subset $X \subseteq \{X_0, \ldots, X_k\}$ of the proposition variables by

$$\psi_j^X \equiv \vartheta(n_j) \wedge \bigwedge_{(n_j, n_l) \in I} (\zeta(n_j, n_l) \to \delta_l^X) \wedge \bigwedge_{(n_j, n_l) \in S} \bigcirc \delta_l^X,$$

$$\delta_j^X \equiv \begin{cases} X_j & \text{if } X_j \in X \\ (vX_j\, \psi_j^{X \cup \{X_j\}}) & \text{otherwise} \end{cases}.$$

Now, define $\varphi^T := \delta_0^\varnothing \equiv vX_0\, \psi_0^{\{X_0\}}$. Induction on $\{X_0, \ldots, X_k\} \setminus X$ shows that all formulas ψ_i^X and δ_i^X are well-defined pre-DSFs whose free proposition variables are among the variables in X. In particular, φ^T is a DSF. $\Diamond$

By standard automata-theoretic constructions, it is easy to see that for every DSF automaton T there is a regular, monotonic, normal and complete DSF automaton T' that accepts exactly the same temporal structures as T, and whose size is linear in the size of T (the construction of a normal DSF automaton T' is linear only if all edge labels of T are in conjunctive normal form, but this is not practically relevant since edge labels will usually be simple).

3 Program Clauses

DSF automata provide an operational representation of DSFs if we interpret edge and node labels as tests to be performed, and assignments of (boolean) values to proposition letters, respectively. A reactive system, however, may change the values of system-owned state components only, whereas environment-owned components may only be read. Formally, distinguish two sets V_E and V_Π of proposition letters that model, respectively, environment and system owned state components, and restrict proposition letters $p \in V_E$ to appear only in antecedents of program clauses. Proposition letters from V_Π may appear both in antecedents and pre-clause bodies. However, to rule out circular definitions where the values of two system owned state components may mutually depend on one another, the programmer will be required to provide a partial order $<$ on the set V_Π. This *dependency order* is then extended to arbitrary literals $L_1 \equiv (\neg)p$, $L_2 \equiv (\neg)q$ by

$$L_1 < L_2 \quad \text{if} \quad (p, q \in V_\Pi \text{ and } p < q) \text{ or } (p \in V_E \cup \{\text{true, false}\} \text{ and } q \in V_\Pi).$$

The following definition defines *program clauses* as the subset of DSFs such that $p < q$ holds whenever the value that will be assigned to q in some state may depend on the value of p in that same state. This additional condition simplifies the interpreter algorithm. In the first-order case, it will be essential to ensure that the construction of each single state in the model construction algorithm will actually terminate.

Definition 6
- *Every literal L is an antecedent.* $read(L) = \{p\}$ *if* $L \equiv (\neg)p$ *for some* $p \in V_E \cup V_\Pi$, $read(L) = \emptyset$ *for* $L \in \{\text{true, false}\}$.
- *If A and B are antecedents, then so are* $A \wedge B$ *and* $A \vee B$. $read(A \wedge B) = read(A \vee B) = read(A) \cup read(B)$.
- *For every proposition letter* $p \in V_\Pi$, p *and* $\neg p$ *are pre-PCs.* $write(p) = write(\neg p) = \{p\}$.
- *If P, Q are pre-PCs and X is a proposition variable then* $P \wedge Q$, $\bigcirc P$, $\bigcirc X$ *and* $(\nu X P)$ *are pre-PCs where* $write(P \wedge Q) = write(P) \cup write(Q)$, $write(\bigcirc P) = write(\bigcirc X) = \emptyset$ *and* $write(\nu X P) = write(P)$.
- *If A is an antecedent and P is a pre-PC such that* $p < q$ *holds for all* $p \in read(A)$, $q \in write(P)$, *then* $A \rightarrow P$ *is a pre-PC.* $write(A \rightarrow P) = write(P)$.
- *Every closed pre-PC (that does not contain free proposition variables) is a program clause (PC).*

The interpreter algorithm of Figure 5.2 attempts to construct a model for a set Φ of program clauses by incremental calculation of the markings of normal DSF automata associated with the clauses in Φ, starting from the environment-defined values of proposition letters $p \in V_E$, and defining the valuation of $q \in V_\Pi$ according to the node labels. Since no backtracking occurs, the interpreter is suitable for the interactive execution of reactive programs. It is easy to see that each node of every DSF automaton is visited at most once, and each edge is tested at most once per iteration of the algorithm. Therefore, both the space and time complexity of the construction of any single state are linear in the size of the normal DSF automata representing the program Φ. By theorem 5, that size is itself linear in the length of the formulas in Φ, provided that all antecedents are in conjunctive normal form.

Example 7 Figure 5.3 illustrates the operation of the algorithm by example of a program that monitors an environment-owned boolean variable *in*, counts (modulo 2) the number of times *in* has changed from *ff* to *tt*, and displays this count in the system-owned boolean variable *out* (where 0 and 1 represent *ff* and *tt*, respectively). *lastin* is an auxiliary (internal) boolean variable containing the value of *in* at the previous state.

Let $\Phi = \{\varphi^1, \ldots, \varphi^q\}$ be a set of temporal program clauses, and $\varepsilon = (\varepsilon_0, \varepsilon_1, \ldots)$ be a temporal structure assigning values $\varepsilon_i(p) \in \{tt, ff\}$ to the environment state components $p \in V_E$. Let $T^k = (N^k, I^k, S^k, \vartheta^k, \zeta^k, n_0^k)$ be a normal DSF automaton associated with the program clause φ^k. Construct the temporal structure $\sigma(\Phi, \varepsilon) = (\sigma_0, \sigma_1, \ldots)$ as follows:

A *configuration* of $T^1, \ldots, T^q$ consists of a q-tuple $\kappa = (v^1, \ldots, v^q)$ of sets $v^k \subseteq N^k$ of nodes of the DSF automata, and a set $\mathcal{E}$ of internal edges that have yet to be inspected. Let "$L \in \zeta^k(n, n')$" denote that the literal L appears as a disjunct in the edge label $\zeta^k(n, n')$. The algorithm defines a sequence of configurations $(\kappa_i^0, \kappa_i^1, \ldots)$ for every state σ_i of the temporal structure $\sigma(\Phi, \varepsilon)$ under construction:

Initialization: Let $i = 0, j = 0$, and define the initial configuration $\kappa_0^0 := (\{n_0^1\}, \ldots, \{n_0^q\})$.
State construction: For the initial configuration $\kappa_i^0 = (v^1, \ldots, v^q)$, let

$$\mathcal{E} := \bigcup_{k=1}^{q} \{(n, n', L) \mid n \in v^k, (n, n') \in I^k, L \in \zeta^k(n, n')\}$$

be the initial set of edges to explore. Define

$$\sigma_i(p) := \left\{ \begin{array}{ll} \varepsilon_i(p) & \text{if } p \in \mathcal{P}_E \text{ is an environment component} \\ tt & \text{if } p \in \vartheta^k(n) \text{ for some } n \in v^k \quad (1 \le k \le q) \\ ff & \text{it } \neg p \in \vartheta^k(n) \text{ for some } n \in v^k \quad (1 \le k \le q) \\ \bot & \text{else} \end{array} \right\}.$$

Complete the construction of σ_i as follows:
1. While $\mathcal{E} \ne \varnothing$, iterate the following steps:
   (a) Set $j := j + 1$. Choose some $(n_k, n_k', L) \in \mathcal{E}$ (where $n_k, n_k' \in N^k$) such that for no $(n_l, n_l', L') \in \mathcal{E}$ it holds that $L' < L$ w.r.t. the dependency order $<$ on literals. If $L \equiv (\neg)p$ and $\sigma_i(p) = \bot$, then arbitrarily define $\sigma_i(p) := tt$ or $\sigma_i(p) := ff$.
   (b) If $L \equiv p$ and $\sigma_i(p) = tt$, or if $L \equiv \neg p$ and $\sigma_i(p) = ff$, or if $L \equiv$ true, then $\kappa_i^j := (v^1, \ldots, \bar{v}^k, \ldots, v^q)$ where $\bar{v}^k = v^k \cup \{n_k'\}$. In this case, define $\sigma_i(p) := tt$ for every $p \in \vartheta^k(n_k')$, and $\sigma_i(p) := ff$ for every p such that $\neg p \in \vartheta^k(n_k')$. Set

$$\mathcal{E} := (\mathcal{E} \setminus \{(n_k'', n_k', L')\}) \\ \cup \{(n_k', n_k'', L') \mid (n_k', n_k'') \in I^k, n_k'' \notin \bar{v}^k, L' \in \zeta^k(n_k', n_k'')\}$$

   Else, $\kappa_i^j := \kappa_i^{j-1}$, and $\mathcal{E} := \mathcal{E} \setminus \{(n_k, n_k', L)\}$.
2. For all $p \in V_\Pi$ such that $\sigma_i(p) = \bot$, arbitrarily define $\sigma_i(p) := tt$ or $\sigma_i(p) := ff$.
3. Let $\kappa_i^j = (v^1, \ldots, v^q)$ be the current configuration. Set $\kappa_{i+1}^0 := (\bar{v}^1, \ldots, \bar{v}^q)$ where $\bar{v}^k = \bigcup_{n \in v^k} S(n)$. Set $i := i + 1$ and $j := 0$, and repeat the state construction.

Fig. 5.2. Interpreter algorithm for propositional program clauses

The definition of the dependency order is irrelevant for this example. DSF automata representing the program clauses as well as the first ten states of an example run are shown as well. The program uses the temporal operators

$$\Box F \equiv \nu X . F \wedge \bigcirc X$$
$$F \text{ upto } G \equiv \nu X F \wedge (\neg G \rightarrow \bigcirc X)$$

(1) ¬*lastin*

(2) □((*in* → ◯*lastin*) ∧ (¬*in* → ◯¬*lastin*))

(3) ¬*out* upto (*in* ∧ ¬*lastin*)

(4) □(*out* ∧ *in* ∧ ¬*lastin* → ◯(¬*out* upto *in* ∧ ¬*lastin*))

(5) □(¬*out* ∧ *in* ∧ ¬*lastin* → ◯(*out* upto (*in* ∧ ¬*lastin*)))

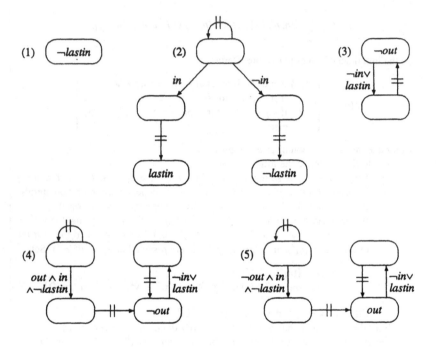

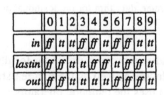

Fig. 5.3. Counter modulo 2 (see example 7)

4 Conflict automata

Obviously, the interpreter can construct a model for a given TL program Φ only if Φ is satisfiable—a nontrivial problem, as shown in [22, 26]. In fact, a stronger condition is required, since the valuation of the environment state components is left unchanged by the interpreter: For a given environment valuation ε, there has to be a model of Φ that agrees with ε on V_E. Fortunately, we can use the automata-theoretic representation of Φ by DSF automata to generate necessary and sufficient conditions (i.e., environment assumptions) for the model construction to succeed. The essential definition is that of a product automaton for DSF automata.

Definition 8 *For regular, normal DSF automata* $T^1 = (N^1, I^1, S^1, \vartheta^1, \zeta^1, n_0^1)$ *and* $T^2 = (N^2, I^2, S^2, \vartheta^2, \zeta^2, n_0^2)$, *the conflict automaton* $T = (N, I, S, \vartheta, \zeta, n_0)$ *is defined as follows:*

$N \subseteq N^1 \times N^2$ is the smallest set of nodes satisfying the following conditions:

- $n_0 := (n_0^1, n_0^2) \in N$
- If $(n^1, n^2) \in N$ and $I^1(n^1) \neq \emptyset$, then $I^1(n^1) \times \{n^2\} \subseteq N$, $I(n^1, n^2) = I^1(n^1) \times \{n^2\}$, $S(n^1, n^2) = \emptyset$, and $\zeta((n^1, n^2), (n, n^2)) = \zeta^1(n^1, n)$ for every $(n, n^2) \in I(n^1, n^2)$.
- If $(n^1, n^2) \in N$ and $I^1(n^1) = \emptyset$, but $I^2(n^2) \neq \emptyset$, then $\{n^1\} \times I^2(n^2) \subseteq N$, $I(n^1, n^2) = \{n^1\} \times I^2(n^2)$, $S(n^1, n^2) = \emptyset$, and $\zeta((n^1, n^2), (n^1, n)) = \zeta^2(n^2, n)$ for every $(n^1, n) \in I(n^1, n^2)$.
- If $(n^1, n^2) \in N$ and $I^1(n^1) = I^2(n^2) = \emptyset$, then $S^1(n^1) \times S^2(n^2) \subseteq N$, $I(n^1, n^2) = \emptyset$ and $S(n^1, n^2) = S^1(n^1) \times S^2(n^2)$.

The node labelling ϑ of T is computed as follows: Let all nodes $(n^1, n^2) \in N$ be unmarked, and let initially $\vartheta(n^1, n^2) := \emptyset$. While there exists some unmarked node $(n^1, n^2) \in N$ such that $\vartheta^1(n^1) \cup \vartheta^2(n^2) \cup \vartheta(n^1, n^2)$ contains either false or a complementary pair $p, \neg p$ of literals:

- Set $\vartheta(n^1, n^2) := \vartheta(n^1, n^2) \cup \{\text{false}\}$.
- For all $((n', n''), (n^1, n^2)) \in I$ where $\zeta((n', n''), (n^1, n^2)) = L_1 \vee \ldots \vee L_k$, set $\vartheta(n', n'') := \vartheta(n', n'') \cup \{\sim L_1, \ldots, \sim L_k\}$ where $\sim L_i$ is $\neg p$ if $L_i \equiv p$, $\sim L_i$ is p if $L_i \equiv \neg p$, $\sim$true = false, and $\sim$false = true.
- For all $((n', n''), (n^1, n^2)) \in S$, set $\vartheta(n', n'') := \vartheta(n', n'') \cup \{\text{false}\}$.
- Mark the node (n^1, n^2) as processed.

Nodes marked in the computation of ϑ as well as those all of whose $(I \cup S)^*$-successors have an empty node label, except the initial node, may be deleted from T. ◊

Example 9 Figure 5.4 illustrates the construction of definition 8 for two regular, normal and monotonic DSF automata corresponding to the DSFs

$$\varphi_1 \equiv r_1 \rightarrow (p \wedge q) \text{ unless } r_2$$
$$\equiv r_1 \rightarrow vX \neg r_2 \rightarrow (p \wedge q \wedge \bigcirc X)$$

$$\varphi_2 \equiv (\neg q \text{ atnext } p) \text{ atnext } r_3$$
$$\equiv vY (\neg r_3 \rightarrow \bigcirc Y) \wedge (r_3 \rightarrow vX (p \rightarrow \neg q) \wedge (\neg p \rightarrow \bigcirc X))$$

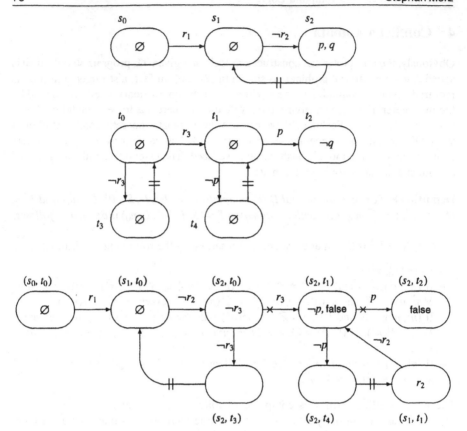

Fig. 5.4. Example for the construction of a conflict graph

Notice how the inconsistency detected at node (s_2, t_2) is propagated back, causing the label of (s_2, t_1) to become inconsistent. After simplification, the DSF associated with the conflict automaton by lemma 5 becomes

$$r_1 \rightarrow vX\,(\neg r_2 \rightarrow \neg r_3 \wedge \bigcirc X) \;\equiv\; r_1 \rightarrow \neg r_3 \text{ unless } r_2.$$

The two following lemmas express that conflict automata characterize the conditions for successful model construction:

Lemma 10 *Let T^1 and T^2 be two regular and normal DSF automata, and let T be their conflict automaton. Every temporal structure accepted by both T^1 and T^2 is also accepted by T.*

Proof: Let σ be a temporal structure, and denote by $M^1 = (m_0^1, m_1^1, \ldots)$, $M^2 = (m_0^2, m_1^2, \ldots)$, and $M = (m_0, m_1, \ldots)$, the markings over σ of T^1, T^2, and T, respectively. It is easy to prove that, for all nodes $n^1 \in N^1$, $n^2 \in N^2$ and every $i \geq 0$, it holds that $(n^1, n^2) \in m_i$ implies $n^1 \in m_i^1$ and $n^2 \in m_i^2$.

Using this auxiliary assertion and the assumption that σ is accepted by both T^1 and T^2, one proves by induction on the definition of the node labelling ϑ of T that for all nodes $(n^1, n^2) \in N$ and every $i \geq 0$ it holds that $(n^1, n^2) \in m_i$ implies $\sigma_i \models \vartheta(n^1, n^2)$. $\Diamond$

Lemma 11 *Let T^1 and T^2 be two regular, normal, complete and monotonic DSF automata, and let T be the conflict automaton of T^1 and T^2. Let σ be some temporal structure that is accepted by T, and denote by $M^1 = (m_0^1, m_1^1, \ldots)$ and $M^2 = (m_0^2, m_1^2, \ldots)$ the markings of T^1 and T^2, respectively, for σ. Then for no $i \geq 0$ there are nodes $n^1 \in m_i^1$, $n^2 \in m_i^2$ such that $\vartheta^1(n^1) \cup \vartheta^2(n^2)$ contains a complementary pair $p, \neg p$ of literals.*
Proof: For a complete DSF automaton T, a node $n_1 \in N$, and a state σ_i of a temporal structure, call $\pi = (n_1, n_2, \ldots, n_k)$ a fullpath for n_1 and σ_i in T if π is a maximal I^*-path in T starting in n_1 such that all edge labels along the path are satisfied in σ_i. Since T is complete, it follows that $I(n_k) = \varnothing$. Let T^1 and T^2 be regular, normal, and complete DSF automata, T their conflict automaton, and σ any temporal structure. It is easy to prove that for every $i \geq 0$, whenever $\pi^1 = (n_1^1, \ldots, n_k^1)$ and $\pi^2 = (n_1^2, \ldots, n_k^2)$ are two fullpaths for n_1^1 and n_1^2 and σ_i in T^1 and T^2, respectively, such that $n_1^1 \in m_i^1$ and $n_1^2 \in m_i^2$, then for some u, v we must have $(n_u^1, n_v^2) \in m_i$ (where $M = (m_0, m_1, \ldots)$ denotes the marking of T over σ). Since T is accepted by σ, it follows that false $\notin \vartheta(n_u^1, n_v^2)$, hence $\vartheta^1(n_u^1) \cup \vartheta^2(n_v^2)$ cannot contain a complementary pair of literals. Since T^1 and T^2 were also assumed to be monotonic, the same follows for $\vartheta^1(n^1) \cup \vartheta^2(n^2)$. $\Diamond$

Given two regular, normal, complete, and monotonic DSF automata T^1 and T^2 for program clauses φ^1 and φ^2, the two preceding lemmas imply that the conflict automaton of T^1 and T^2 characterizes a necessary and sufficient condition for the model construction for φ^1 and φ^2 to succeed. In the more general case of a set Φ of program clauses, it is easy to see that conflict automata for every pair of program clauses of Φ, including two distinct copies for every single program clause, precisely identify the conditions for runs of Φ to succeed. This fact is due to the absence of disjunctions in pre-clause bodies (and hence in node labels), i.e. to the high degree of determinism required of program clauses. These ideas are formally captured by the following definition of *compatibility* that allows for environment assumptions, as is appropriate for a rely-guarantee style of program development.

Definition 12 *Let $\Phi = \{\varphi_1, \ldots, \varphi_q\}$ be a set of program clauses, and Ψ be some closed vTL formula over $\mathcal{V}_E$. Denote by $T_1, \ldots, T_q$ regular, normal, monotonic, and complete DSF automata corresponding to $\varphi_1, \ldots, \varphi_q$. Φ is called Ψ-compatible if for all $1 \leq k \leq q$ and $1 \leq j \leq k$, the DSFs corresponding to the conflict automata of T_j and T_k follow from Ψ and $\{\varphi_1, \ldots, \varphi_{k-1}\}$.*

By an induction on the number of steps taken by the interpreter algorithm, the following theorem has been shown in [18]:

Theorem 13 *Let Φ and Ψ be as in definition 12.*
(i) *Φ is Ψ-compatible iff for every model $\varepsilon = (\varepsilon_0, \varepsilon_1, \ldots)$ of Ψ there is a model of Φ that agrees with ε on $\mathcal{V}_E$.*
(ii) *If Φ is Ψ-compatible and ε is a model of Ψ, then the interpreter algorithm of Figure 5.2 constructs a model σ of Φ that agrees with ε on $\mathcal{V}_E$.*

Notice that the property of being Ψ-compatible w.r.t. some given environment assumption Ψ can be decided "statically" by an analysis of the given program Φ. For the example program of Figure 5.3, all conflict automata turn out to be trivial, hence the program is Ψ-compatible w.r.t. any environment condition Ψ.

5   First-order program clauses

Propositional TL can describe systems of bounded state space. When it is impossible or inconvenient to bound the state space a priori, first-order TL offers the appropriate expressiveness. Fortunately, most notions introduced above generalize readily to the first-order case. First-order TL can represent local state using either *flexible variables*, which correspond to program variables of conventional imperative programming languages, or *flexible predicates*, which are appropriate to model sets of data such as database relations. On the other hand, I assume all function (and constant) symbols, as well as equality to be rigid.

The universe and the interpretation of the function symbols are provided by an underlying first-order model. In terms of an actual interpreter realization, this means that they will be implemented in a suitable host programming language. Every state of a temporal structure interprets the predicate symbols and the flexible variables of the language. First-order νTL also contains a set of *rigid variables* [15] to bind data values, as in classical first-order logic. I will allow quantification over rigid variables only. For a term t, let $FRV(t)$ and $FFV(t)$ denote, respectively, the sets of rigid and flexible variables occurring in t, and similarly for a formula F.

The overall structure of first-order DSFs and program clauses is the same as in the propositional case. As is conventional in logical and rule-based programming languages, all rigid variables that occur in clauses are interpreted as being in the scope of an implicit universal quantifier. This ensures that all program clauses express safety properties, using the following fact [18] that extends Fact 1:

Fact 14 *Let F be a first-order νTL formula such that no subformula (νXG) of F appears either negatively or in the scope of a quantifier of existential force. Then F defines a safety property.*

For first-order TL programs that contain flexible predicate symbols, it has to be ensured that the predicate extensions at each state remain finite; otherwise the model could not be effectively presented. This requirement entails some conditions on the rigid variables appearing in program clauses, similar to the definition of safe queries in database theory [25]. Since the universe will in general be (conceptually) infinite, it follows that the complements of predicate extensions will not be explicitly representable. Hence, the interpreter will adopt a "minimal-model" approach during the state construction: For an n-ary predicate P and arbitrary elements $d_1, \ldots, d_n$ of the universe, $(d_1, \ldots, d_n) \in \sigma_i(P)$, for any state σ_i, will hold only if this follows from an instance of some program clause. This convention eliminates the need for negative literals to appear in pre-clauses. Negative literals may, however, appear in antecedents, where they are interpreted as tests.

Formally, first-order DSFs are defined as follows:

Definition 15

- *For any term t, define the set Bind(t) by Bind(t) = {y} if t ≡ y is a rigid variable, and Bind(t) = ∅ else. Intuitively, Bind(t) denotes the set of rigid subject variables for which t may establish a binding if it appears in an antecedent.*
- *Atomic formulas $P(t_1, \dots, t_m)$ or $t = t'$ as well as their negations and the propositional constants* true *and* false *are literals. For a literal L, the set Bind(L) is defined by*

$$Bind(P(t_1, \dots, t_m)) = (\bigcup_{i=1}^{m} Bind(t_i)) \setminus (\bigcup_{Bind(t_j)=\emptyset} FRV(t_j)),$$
$$Bind(t = t') = Bind(t) \setminus FRV(t'),$$
$$Bind(\neg P(t_1, \dots, t_m)) = \emptyset, \qquad Bind(t \neq t') = \emptyset,$$
$$Bind(\text{true}) = \emptyset, \qquad Bind(\text{false}) = \emptyset.$$

- *Every literal is an antecedent. If A, B are antecedents then $A \wedge B$ and $A \vee B$ are also antecedents where $Bind(A \wedge B) = Bind(A \vee B) = Bind(A) \cap Bind(B)$.*
- *Atomic formulas of the form $P(t_1, \dots, t_m)$ or $t = t'$ are pre-DSFs. Define the set $Open(P(t_1, \dots, t_m)) = \bigcup_{i=1}^{m} FRV(t_i)$, and $Open(t = t') = FRV(t) \cup FRV(t')$. Intuitively, Open(P) denotes the set of rigid variables that have to be bound before the pre-DSF P is evaluated.*
- *If A is an antecedent, X is a proposition variable and F, G are pre-DSFs then $F \wedge G$, $\bigcirc F$, $\bigcirc X$, $A \rightarrow F$ and $(vX\ F)$ are pre-DSFs where $Open(F \wedge G) = Open(F) \cup Open(G)$, $Open(\bigcirc F) = Open(vX\ F) = Open(F)$, $Open(\bigcirc X) = \emptyset$ and $Open(A \rightarrow F) = (FRV(A) \cup Open(F)) \setminus Bind(A)$.*
- *If F is a pre-DSF that does not contain any free proposition variable and where $Open(F) = \emptyset$, then F is a deterministic safety formula (DSF).*

First-order DSFs may be represented in an automata-theoretic framework similar to the propositional DSF automata introduced above. The only extension is to record the set of rigid variables bound at each node and to impose variable conditions on literals that appear in node and edge labels, corresponding to those for DSFs. First-order program clauses are then again defined by distinguishing between state components owned by the environment and those owned by the system. In particular, let V_E, V_Π, P_E and P_Π, respectively, denote the sets of flexible variables and predicate symbols owned by environment and system. Again, the programmer has to provide a partial dependency order $<$ on the set $V_\Pi \cup P_\Pi$ of system-owned state components. Atomic pre-clauses may be either of the form $P(t_1, \dots, t_m)$ for $P \in P_\Pi$ or of the form $v = t$ for $v \in V_\Pi$ (resembling an assignment to the program variable v) provided that $v' < P$ (resp., $v' < v$) holds for every flexible variable v' occurring in any term t_i (resp., in t). The definition of PCs then proceeds inductively as in the propositional case.

The definition of conflict automata to generate formulas that guarantee the success of the interpreter algorithm is in fact easier than in the propositional case: Since negative literals have been ruled out in pre-clause bodies, the only possible inconsistencies may arise from assignments of different values to the same flexible variable. If two nodes n^1 and n^2 of the original DSF automata contain such assignments $v = t$ and $v = t'$, the corresponding node (n^1, n^2) of the product automaton will contain $t = t'$ in its node label.

Having defined conflict automata, the condition of Ψ-compatibility of some program Φ w.r.t. an environment assumption Ψ remains as before. Given a program Φ, all models generated by the interpreter will have minimal extensions of flexible system predicates, and therefore satisfy additional properties over the logical semantics of Φ. To make the logical and operational semantics of Φ agree, a notion of *program completion* is defined, similarly as for negation-free classical logic programs [13], using the automata-theoretic representation of program clauses. The compatibility criterion can then be relativized to the completed program.

The interpreter algorithm is a relatively straightforward extension of the propositional interpreter of Figure 5.2. For first-order programs without flexible predicates, the complexity to build each single state of the structure is still linear in the size of the automata representing the program clauses. However, if flexible predicate symbols are allowed, it is easy to find examples that exhibit exponential growth of the predicate extensions along the state sequence. It may be worthwhile to impose further syntactic conditions to prevent such undesired behaviour.

---

(1) $\Box(Hungry(x) \rightarrow Waiting(x)$ unless $Eat(x) \vee Enough(x))$

(2) $\Box(Waiting(x) \rightarrow (Left(x)$ upto $Reclaim(x \ominus 1) \vee Enough(x))$
        atnext $\neg Right(x \ominus 1))$

(3) $\Box(Left(x) \wedge \neg Left(x \oplus 1) \rightarrow \bigcirc(Right(x) \wedge Eat(x)$ unless $Enough(x)))$

(4) $preferred = 0$ upto $Reclaim(0)$

(5) $\Box(Reclaim(x) \rightarrow \bigcirc((Right(x) \wedge Eat(x)$ unless $Enough(x)) \wedge$
        $(preferred = x \oplus 1$ upto $Reclaim(x \oplus 1))))$

where $Reclaim(x) \equiv preferred = x \wedge Left(x) \wedge Left(x \oplus 1) \wedge \neg Right(x \oplus 1)$,
        $x \oplus 1 \equiv (x + 1) \bmod n$, $x \ominus 1 \equiv (x - 1) \bmod n$

Fig. 5.5. Butler for n philosophers

As an example for a first-order temporal logic program, figure 5.5 shows an implementation of a "butler" module for the dining-philosopher program for an arbitrary number n of philosophers. Here, *Hungry* and *Enough* are environment-owned unary predicate symbols, *Waiting*, *Left*, *Right*, and *Eat* are system-owned unary predicate symbols, and *preferred* is a system-owned flexible variable. The dependency order is defined by $Eat < Waiting < Right < Left$.

6   Discussion

Compared to conventional programming languages for reactive systems, temporal logic programming promises a higher level of abstraction that relieves the developer from explicitly stating the next-state relation. In our approach, it is inferred by the compiler that constructs the DSF automata. Another important feature is the absence of backtracking, because this allows for incremental model construction. As a sanity check on programs, it is possible to automatically generate formulas that guarantee successful runs. The developer should then prove these formulas from given environment assumptions. (In the propositional case, this proof can be automated, since propositional vTL is decidable).

By comparison, the languages proposed in [19] or [24] are basically similar to conventional programming languages, and offer only few higher-level constructs. At the other end of the spectrum, the language METATEM [4] offers the expressiveness of full linear-time temporal logic, and imposes only a syntactic normal form on admissible formulas. Consequently, the interpreter algorithm becomes much more complex, since it needs to detect failure of the model construction, and backtrack if it does. In particular, the environment and the system cannot truly run interactively, because the environment cannot be sure that the system's reactions will remain stable. In general, early detection of failure situations requires theorem-proving abilities and becomes an undecidable problem. Another tradeoff that has been resolved differently concerns the presence of past-time connectives in antecedents. Since these are allowed in METATEM, the interpreter has to store the entire computation history. In contrast, our interpreter can evaluate all antecedents solely on the basis of the present state. On the other hand, the programmer may have to define auxiliary state components that serve as a memory (such as the predicate *Waiting* in the dining philosophers example).

The languages TEMPLOG [1, 6] and CHRONOLOG [20] aim to base a temporal logic programming language on theorem proving in a similar way as PROLOG is based on SLD-resolution. In such approaches, given the strong incompleteness of first-order temporal logics, the class of admissible formulas has to be restricted severely. Also, it is not so clear intuitively how to generate infinite runs of reactive systems from finite proofs than it is to interpret temporal structures as such runs. It has recently become clear [7, 9, 10] that the powerful mechanisms of constraint logic programming languages can be employed to perform temporal logic proofs, thereby obviating the need for special-purpose TL theorem provers.

Acknowledgements

It is a pleasure to express my gratitude toward all those who by constructive criticism have helped to form the ideas or to improve the presentation. In particular, I want to thank Jean Paul Bahsoun, Luis Fariñas del Cerro, Hans Gaßner, Fred Kröger, Frank Leßke, Holger Schlingloff, Corinne Servières, and Martin Wirsing, as well as the other members of the nonclassical logics groups at Munich and Toulouse.

References

1. M. Abadi: *Temporal-Logic Theorem Proving*. Ph.D. thesis, Department of Computer Science, Stanford University (1987)

2. B. Alpern, F. B. Schneider: *Recognizing safety and liveness*. Distributed Computing 2, pp. 117–126 (1987)

3. B. Banieqbal, H. Barringer: *Temporal Logic with Fixed Points*; in: Proc. Temporal Logic in Specification (B. Banieqbal et al., ed.). LNCS 398, pp. 62–74, Springer-Verlag (1989)

4. H. Barringer, M. Fisher, D. Gabbay, G. Gough, R. Owens: METATEM: *A Framework for Programming in Temporal Logic*; in: J. W. de Bakker, W.-P. de Roever, G. Rozenberg (eds.): Stepwise Refinement of Distributed Systems — Models, Formalisms, Correctness. LNCS 430, Springer-Verlag (1990). See also [8].

5. H. Barringer, R. Kuiper, A. Pnueli: *Now You May Compose Temporal Logic Specifications*. Proc. Sixteenth ACM Symposium on the Theory of Computing, pp. 51–63 (1984)

6. M. Baudinet: *Temporal Logic Programming is Complete and Expressive*; in: Proc. ACM Conf. on Princinples of Programming Languages, pp. 267–280 (1989)

7. C. Brzoska: *Temporal Logic Programming with Metric and Past Operators*. Proc. Workshop on Executable Modal and Temporal Logics, in this volume.

8. M. Fisher: *Concurrent* METATEM — *The Language and its Applications*. Proc. Workshop on Executable Modal and Temporal Logics, IJCAI 1993, in this volume.

9. T. Fruehwirth: *Temporal Logic and Annotated Constraint Logic Programming*. Proc. Workshop on Executable Modal and Temporal Logics, IJCAI 1993, in this volume.

10. J. Koehler, R. Treinen: *Constraint Deduction in an Interval-based Temporal Logic*. Proc. Workshop on Executable Modal and Temporal Logics, IJCAI 1993, in this volume.

11. S. Kono: *A Combination of Clausal and Non Clausal Temporal Logic Program*. Proc. Workshop on Executable Modal and Temporal Logics, IJCAI 1993, in this volume.

12. F. Kröger: *Temporal Logic of Programs*. EATCS Monographs on Theoretical Computer Science 8. Berlin: Springer-Verlag (1987)

13. J. W. Lloyd: *Foundations of Logic Programming*. Springer-Verlag (^{2}1987)

14. Z. Manna, A. Pnueli: *A Hierarchy of Temporal Properties*; in: Proc. ACM Symposium on Principles of Distributed Computing (1990)

15. Z. Manna, A. Pnueli: *The Temporal Logic of Reactive and Concurrent Systems — Specification*. New York etc.: Springer-Verlag (1992)

16. Z. Manna, P. Wolper: *Synthesis of Communicating Processes from Temporal Logic Specifications*. ACM Transactions on Programming Languages 6, pp. 68–93 (1984)

17. S. Merz: *Decidability and Incompleteness Results for First-Order Temporal Logics of Linear Time*. Journal of Applied Non-Classical Logics 2(2) (1992)

18. S. Merz: *Temporal Logic as a Programming Language*. Ph.D. thesis, Ludwig-Maximilians-Universität (1992)

19. B. Moszkowski: *Executing Temporal Logic Programs*. Cambridge (UK): Cambridge University Press (1986)
20. M. A. Orgun, W. W. Wadge: *Towards a unified theory of intensional logic programming*. Journal of Logic Programming, **13**(4), pp. 413ff (1992)
21. A. Pnueli: System Specification and Refinement in Temporal Logic. Proc. Conf. on Foundations of Software Technology and Theoretical Computer Science. Springer LNCS 652, pp. 1–38 (1992)
22. A. P. Sistla, E. M. Clarke: *The Complexity of Propositional Linear Temporal Logic*. Journal of the ACM **32**(3), pp. 733–749 (1985)
23. A. Szałas: *A Complete Axiomatic Characterization of First-Order Temporal Logic of Linear Time*. TCS **54**, pp. 199–214 (1987)
24. Z. Tang: *Toward a Unified Logical Basis for Programming Languages*. In: R.E.A. Mason, ed.: Proc. IFIP Congress '83. Amsterdam: Elsevier Publishers B.V. (1983)
25. J. D. Ullmann: *Principles of Database and Knowledge-Base Systems*. Computer Science Press (1988)
26. P. Wolper: *Temporal Logic Can Be More Expressive*. Information and Control **56**, pp. 72–99 (1983)

Towards a Semantics for Concurrent METATEM

Michael Fisher

Abstract. Concurrent METATEM is a programming language based on the notion of concurrent, communicating objects, where each object directly executes a specification given in temporal logic, and communicates with other objects using asynchronous broadcast message-passing. Thus, Concurrent METATEM represents a combination of the direct execution of temporal specifications, together with a novel model of concurrent computation. In contrast to the notions of predicates as processes and stream parallelism seen in concurrent logic languages, Concurrent METATEM represents a more coarse-grained approach, where an object consists of a set of logical rules and communication is achieved by the evaluation of certain types of predicate. Representing concurrent systems as groups of such objects provides a powerful tool for modelling complex reactive systems. Being based upon executable temporal logic, objects in isolation have an intuitive semantics. However, the addition of both operational constraints upon the object's execution and global constraints provided by the model of concurrency and communication, complicates the overall semantics of networks of objects. It is this, more complex, semantics that we address here, where a basis for the full semantics of Concurrent METATEM is provided.

1 Introduction

A wide variety of computer systems are *reactive*. A reactive system does not simply read in a set of inputs and produce, on termination, a set of outputs [22, 16]. Reactive systems are typically concurrent or distributed, and contain elements that are constantly *reacting* to stimuli from their environment. In this paper, we provide a semantics for Concurrent METATEM, a language for representing and implementing a subclass of reactive systems [14]. This subclass contains reactive systems that are concurrent and whose elements represent self-contained entities that communicate through message-passing. These entities encapsulate both data and behaviour and, hence, can be termed *objects*.

Such reactive systems are sometimes termed *concurrent object-based systems* [28].

Concurrent METATEM has been developed from the sequential execution of temporal specifications provided by METATEM, an executable temporal logic described in [4, 11]. Thus, individual objects execute temporal specifications and communicate with their environment at certain times by *broadcasting* information.

Adopting this style of language provides the programmer/designer with a flexible way to view concurrent systems, where computation is carried out within groups of objects broadcasting, listening and executing asynchronously. Such executing objects are generally coarse-grained, with an object consisting of a set of logical rules which define constraints upon certain predicates. The logic we use within each individual object is powerful, and has been shown to be useful in describing and implementing various properties of reactive systems [20, 4].

The structure of this paper is as follows. In §2, we outline the main features of Concurrent METATEM, including the internal representation and execution of objects using temporal logic, and the interaction of sets of objects using concurrency and communication. In addition to providing a description of the basic Concurrent METATEM system, we present a range of extensions that have either been proposed or implemented. In §3, we provide a semantics for a basic Concurrent METATEM system using a *dense* temporal logic. Although Concurrent METATEM is based on the execution of a *discrete* temporal logic, we present its semantics in terms of dense temporal logic in order to realistically model the asynchronous execution of individual objects. We then show how this semantics can be refined in order to represent some of the extensions described earlier. Finally, in §4, we present our conclusions together with an outline of future work on the semantics of Concurrent METATEM.

2 Concurrent METATEM

In this section we give an outline of Concurrent METATEM. (The reader is referred to [14, 10] for more details.) We begin by motivating and describing the overall computational model, follow this by examining the internal definition and execution of objects within this model, and conclude with a brief discussion of extensions to the basic language. As the internal definition of Concurrent METATEM objects is based on temporal logic, we also include a brief overview of the logic we use, together with an outline of the execution model for temporal logic formulae provided by METATEM [4, 11].

2.1 The Computational Metaphor

The abstract model used in Concurrent METATEM combines the two notions of *objects* and *concurrency*. Objects are here considered to be self contained entities, encapsulating both data and behaviour, and communicating via message-passing.

The predominant approach to concurrent object-based systems is that of the *actor* model of computation [17, 1]. This model has not only been implemented directly [19], but has also been used as the basis for combining objects and concurrency in other languages, such as ABCL [27] and Concurrent Prolog [24, 18]. The metaphor used in the actor model is of a mail system, with messages being addressed and sent directly to individual actors. Thus, actor systems are based upon point-to-point message-passing and message-driven computation.

Concurrent METATEM is based on a rather different computational metaphor[1]. Although objects are concurrently executing entities communicating through message-passing, they have the following fundamental properties which contrast, for example, with the actor model.

1. The basic method of communication between objects is *broadcast* message-passing.
2. Objects are not message driven — they begin executing from the moment they are created and continue even while no messages are received.
3. An object can change its interface (i.e., the messages that it recognises) dynamically.

So, rather than seeing computation as objects sending mail messages to each other, and thus invoking some activity, computation in Concurrent METATEM can be visualised as independent objects *listening* to messages broadcast from other objects.

Although reliable broadcast message-passing may be expensive to implement (but see [6]), and secure communication may be difficult to achieve, using broadcast as the primitive message-passing mechanism provides several advantages over point-to-point message-passing. For example, this model of communication affords the opportunity for higher reliability and availability within the system [3, 2]. The use of the broadcast mechanism provides robustness in that 'duplicate' objects can be easily represented within Concurrent METATEM. These objects can either provide an alternative source for replies to messages or can be used as 'reserve' objects that can take the place of the primary object if its processor crashes. For example, replication mechanisms such as those described in [8] (based on objects which mirror and monitor the primary object in case of disruption) and [25] (based on *opportunistic recovery* from processor crashes) may be represented and implemented within Concurrent METATEM. A further advantage is that Concurrent METATEM may be used in systems where the individual objects do not necessarily know the names/addresses of other objects.

Thus, a system consists of a set of concurrently executing objects which communicate through asynchronous broadcast message-passing. Objects are the basic computational entities in a system and each object has two components:

1. an abstract *interface definition*, and,
2. an *internal definition*.

The interface definition is abstract in that the same definition can be used regardless of the object's internal definition. Although we use executable temporal logic to implement individual objects, this opens up the possibility of developing *heterogeneous* systems consisting of objects implemented in a variety of ways.

---

[1] Elements of this model of computation have appeared in a variety of guises, from operating systems [7] to A.I. [21], and so we do not claim that it is new — only that it is a useful computational model for executable logics.

Object Interfaces Networks of objects communicate via broadcasting messages and individual objects only act upon certain identified messages. Thus, an object must be able to filter out messages that it wishes to recognise, ignoring all others. The definition of which messages an object recognises, together with a definition of the messages that an object may itself produce, is provided by the *interface definition* for that particular object. The interface definition for an object, for example 'stack', is defined in the following way

$$\texttt{stack(pop,push)[popped,full]}.$$

Here, {pop, push} is the set of messages the object recognises, while {popped, full} is the set of messages the object might produce itself. Note that these sets need not be disjoint – an object may broadcast messages that it also recognises. In this case, messages sent by an object to itself are recognised immediately. Note also that many different objects may broadcast and recognise the same messages.

Given this general model for concurrent systems, we choose to represent and implement individual objects using an executable temporal logic. Thus, we next give a brief, and relatively informal, introduction to temporal logic followed, in §2.3, by an outline of the execution mechanism used for temporal formulae.

2.2 Temporal Logic

Temporal logic can be seen as classical logic extended with various modalities representing temporal aspects of logical formulae [9]. The temporal logic we use is based on a linear, discrete model of time. Thus, time is modelled as an infinite sequence of discrete states, with an identified starting point, called 'the beginning of time'. Classical formulae are used to represent constraints within individual states, while temporal formulae represent constraints *between* states. As formulae are interpreted at particular states in this sequence, operators which refer to both the past and future are required.

The future-time temporal operators used in this paper are as follows.

- The *sometime in the future* operator, '$\Diamond$'.
    $\Diamond \varphi$ is true now if φ is true *sometime* in the future.
- The *always in the future* operator, '$\Box$'.
    $\Box \varphi$ is true now if φ is true *always* in the future.
- The *until* operator, '$\mathcal{U}$'.
    $\varphi \mathcal{U} \psi$ is true now if ψ is true from now *until* a moment in the future when ψ becomes true.

Similarly, connectives are provided to enable formulae to refer to the *past*. The past-time temporal operators needed for the examples below can be described as follows.

- The *since* operator, '$\mathcal{S}$'.
    $\varphi \mathcal{S} \psi$ is true now if ψ was true in the past and φ was true from that moment until (but not including) the present moment.
- The *sometime in the past* operator, '$\blacklozenge$', which is the past-time analogue of '$\Diamond$'.

- The *always in the past* operator, '■', which is the past-time analogue of ' □ '.
- The *beginning of time* operator, '**start**'.

   start is only true at the beginning of time.
- The *strong last-time operator*, '**O**'.

   O φ is true if there was a last moment in time and, at the moment, φ was true.

There are many other temporal operators used in temporal logics in general, and in METATEM in particular, though they will not be mentioned here, not only for clarity but also since such temporal formulae can be reduced to a normal form.

A Normal Form for Temporal Formulae Fortunately, although a range of temporal operators can be defined, we are able to transform arbitrary temporal formulae into a normal form called SNF [13] which basically consists of a set of 'rules' each of one of the following forms:

$$\textbf{start} \Rightarrow \bigvee_{j=1}^{r} m_j \qquad (\text{an } \textit{initial } \Box\text{-rule})$$

$$\textbf{O} \bigwedge_{i=1}^{q} k_i \Rightarrow \bigvee_{j=1}^{r} m_j \qquad (\text{a } \textit{global } \Box\text{-rule})$$

$$\textbf{start} \Rightarrow \Diamond l \qquad (\text{an } \textit{initial } \Diamond\text{-rule})$$

$$\textbf{O} \bigwedge_{i=1}^{q} k_i \Rightarrow \Diamond l \qquad (\text{a } \textit{global } \Diamond\text{-rule})$$

where each k_i, m_j or l is a literal. Note that the left-hand side of each initial rule is a constraint only on the *first* state, while the left-hand side of each global rule represents a constraint upon the previous state. The right-hand side of each $\Box$-rule is simply a disjunction of literals referring to the current state, while the right-hand side of each $\Diamond$-rule is a single eventuality (i.e., '$\Diamond$' applied to a literal).

It should be noted that the use of temporal logic as the basis for the computation rules gives an extra level of expressive power over the corresponding classical logics. In particular, operators such as '$\Diamond$' give us the opportunity to specify future-time (temporal) indeterminacy. Representation in SNF allows us to capture these expressive capabilities concisely.

2.3 METATEM

METATEM is a programming and modelling language based on temporal logic. It uses a set of 'rules', couched in temporal logic, to represent an object's internal definition [4]. These rules are generally of the form [15]

   'past formula' **implies** 'present or future formula'

Such rules are applied at every moment in time (i.e., at every step of the execution). In our particular case, the rules are in SNF.

In executing a set of SNF rules, the aim is to produce a model for those rules, but to do so using a *forward-chaining* process. Thus, execution in METATEM follows a cycle of checking if the left-hand side of each rule is satisfied in the current execution and, if it is, ensuring that the right-hand side is also satisfied.

The operator used to represent the basic temporal indeterminacy within METATEM is the *sometime* operator, '$\Diamond$'. When a formula such as $\Diamond\varphi$ is executed, the system must try to ensure that φ *eventually* becomes true[2]. As such eventualities might not be able to be satisfied straight away, we must keep a record of the unsatisfied eventualities, retrying them as execution proceeds [11].

As an example of a simple METATEM program, consider the following set of rules. (Note that these rules are not meant to form a meaningful program – they are only given for illustrative purposes.)

$$\textbf{start} \Rightarrow \text{popped}$$
$$\bigcirc\text{pop} \Rightarrow \Diamond\text{popped}$$
$$\bigcirc\text{push} \Rightarrow \text{full} \lor \text{popped}$$

Looking at these program rules, we see that popped is made true at the beginning of time and whenever pop is true in the last moment in time, a commitment to eventually make popped true is given. Similarly, whenever push is true in the last moment in time, then either full or popped must be made true.

For a more detailed description of the METATEM execution mechanism, see [11].

2.4 Concurrent Operational Model

Given processes executing METATEM as above, we develop an operational model for their concurrent execution and communication as follows. This develops and extends METATEM into Concurrent METATEM.

The basic predicates used in the logic are categorised as follows, with several categories of predicate corresponding to messages to and from the object.

- *Environment* predicates, which represent incoming messages.
  An environment predicate can be made true if, and only if, the corresponding message has just been received. Thus, a formula containing an environment predicate, such as 'push', is only true if a message of the form 'push' has just been received.
- *Component* predicates, which represent messages broadcast from the object.
  When a component predicate is made true, it has the (side-)effect of broadcasting the corresponding message to the environment. For example, if the formula 'popped' is made true, where popped is a component predicate, then the message 'popped' is broadcast.
- *Internal* predicates, which have no external effect.
  These predicates are used as part of formulae participating in the internal computation of the object and, as such, do not correspond either to message-sending or message reception.

---

[2] Thus, such formulae are often called eventualities.

This category of predicates may include various *primitive* operations.

Once the object has commenced execution, it continually follows a cycle of reading incoming messages, collecting together the rules that 'fire' (i.e., whose left-hand sides are satisfied by the current history), and executing one of the disjuncts represented by the conjunction of right-hand sides of 'fired' rules.

Note that, by default, the message-passing mechanism does not guarantee that the order of arrival of messages is the same as the order of sending.

2.5 Extensions

While the above provides the basic Concurrent METATEM system, we here describe some of the extensions to this that have either been proposed or implemented.

Synchronisation Though objects execute asynchronously, provision for the *synchronisation* of objects is often useful. For example, an object might broadcast a request for something, then continue executing until a reply is received. While it is waiting for the reply, the object can process other messages. Alternatively, the object might broadcast its request, then *suspend* until an appropriate reply is received. In this case, it will not process any other messages in between the request being sent and the reply being received. This can be considered as a process of synchronising the object with the reply message.

Within Concurrent METATEM, both approaches can be represented, though here we will only outline the second one where an object sends out a request, *suspends* waiting for an answer, and performs some action when an answer arrives. If we imagine that ask is a component predicate, while answer is an environment predicate, then we can represent such an object's rules as follows.

$$\ldots \Rightarrow \text{ask}$$
$$\bullet \quad \text{answer} \Rightarrow \text{do\_it}$$
$$\bullet \quad \text{ask} \Rightarrow \text{answer}$$

If an ask message has been sent, then the only way to satisfy this last rule is to ensure that answer is received in the next state. Thus, the object cannot execute further until the required message arrives, and consequently it is suspended. In this way, objects in Concurrent METATEM can synchronise with selected messages.

Groups An extension being developed at present is the provision of structuring using *groups*. Each object may be a member of several groups. When an object sends a message, that message is, by default, broadcast to all the members of its group(s), but to no other objects. Alternatively an object can select to broadcast only to certain groups (of which it is a member). This mechanism allows the development of complex structuring within the object space and provides the potential for more innovative applications [12].

We assume a group is simply represented by a set of the names of the objects that it contains. This provides us with a simple description of groups that will be used in giving their semantics (see §3.5).

Message Queues Each object in the system is associated with a *message queue*, representing the messages that the object has recognised, but has yet to process. The number of messages that an object reads from its message queue during an execution step is initially defined by the object's interface. By default, the execution of a Concurrent METATEM object is based on the set of messages received by the object since the last execution step it completed. Thus, objects consume sets of messages, rather than enforcing some linearisation on the order of arrival of messages.

However, it is sometimes useful to provide alternative message-manipulation behaviour, for example if we require that the object processes just one message at a time. Thus, Concurrent METATEM will provide mechanisms for dynamically changing this behaviour, though as yet these have not been implemented.

Dynamic Interfaces The interface definition of an object defines the initial set of messages that are recognised by that object. However, the object may dynamically change the set of messages that it recognises. In particular, an object can either start 'listening' for a new type of message, or start 'ignoring' previously recognised message types. For example, given an original object interface such as

```
stack(pop,push)[popped,full]
```

the object may dynamically choose to stop recognising 'pop' messages, perhaps by executing 'ignore(pop)'. This effectively gives the object the new interface

```
stack(push)[popped,full].
```

Backtracking In general, if an object's execution mechanism is based on the execution of logical statements, then a computation may involve backtracking. In Concurrent METATEM, objects may backtrack, with the proviso that an object may not backtrack past the broadcasting of a message. Consequently, in broadcasting a message to its environment, an object effectively *commits* the execution to that particular path. Thus, the basic operation of an object can be thought of as a period of internal execution, possibly involving backtracking, followed by appropriate broadcasts to its environment.

3   Semantics for Concurrent METATEM Systems

As one might expect, given that the language is based upon an executable temporal logic, a *temporal* semantics can be given for Concurrent METATEM. However, because both operational constraints and linguistic extensions have been added to the core language, defining the semantics for a Concurrent METATEM system is not as straightforward as it might seem. In particular, when asynchronous execution is introduced, each object effectively constructs a different temporal model. Because of this, we provide the semantics of (asynchronous) Concurrent METATEM using a *dense* temporal logic even though the individual objects each execute formulae of discrete temporal logic.

We will begin by giving a temporal semantics for one of the simplest versions of Concurrent METATEM, i.e., one that encompasses the following constraints.

1. All objects execute in synchrony
   — the execution steps of each object match exactly.
2. Messages broadcast are instantaneously received by every object
   — there is no delay in transmission.
3. Objects may not synchronise on external messages
   — environment predicates are only allowed in the left-hand side of METATEM rules.
4. There is no structuring of the object-space in to groups
   — all objects occupy the same message/object space.
5. Objects can not dynamically change their interface definitions
   — objects can neither listen nor ignore.

We will extend this semantics by successively removing each of the above constraints. Note that a semantics can be given for this basic system using a *discrete* temporal logic. However, when we remove the constraint that all objects should execute synchronously, it is more natural to give the temporal semantics using a dense temporal logic.

Object Structures In order to give a semantics for Concurrent METATEM, we assume a simple syntactic structure for representing objects in the language. For the purpose of this discussion, an object (e.g., *Obj*) is represented by a structure indexed by the following fields.

Rules — the set of rules comprising the object.
Env   — the set of environment predicates.
Comp — the set of component predicates.
Int   — the set of internal predicates.

Thus, for example, we can access the set of component predicates of the object *Obj* via *Obj.Comp*.

Predicate Renaming To avoid any unwanted name clashes between different objects, we rename each predicate, for example p, used in object O by a new predicate p-O. Thus, if we make predicate p true in a particular object, this will not directly affect the truth value of p in a different object. However, we also need to add communication constraints linking certain predicates together. The basic predicates p, q, etc., are used to connect the sender's name for the predicate to the receiver's name for the predicate (see below).

Semantic Definitions In order to simplify the description, we will present a semantics for propositional Concurrent METATEM. This semantics will be given using the semantic function $\mathcal{M}$, with a subscript representing the type of syntactic entity to which this function is applied. For example, $\mathcal{M}_{rule}$ gives the semantics of individual rules, while $\mathcal{M}_{system}$ gives the meaning of the whole system. The types of these semantic functions are obvious, mapping the particular syntactic structure (e.g., rule, object, system) on to a temporal logic formula.

3.1 Basic Semantics

We begin with a simple Concurrent METATEM system encompassing all the constraints described above. We give the semantics of such a system, S, consisting of the Concurrent METATEM objects $O_1, O_2, \ldots, O_n$, as follows.

$$\mathcal{M}_{system}(S) = \bigwedge_{i=1}^{n} \mathcal{M}_{object}(\text{rename}(O_i))$$

$$\mathcal{M}_{object}(O) = \mathcal{M}_{rules}^{O}(O.rules) \wedge \text{comms}(O)$$

where $\text{rename}(O_i)$ carries out the renaming of predicates within object O_i, as described above. The predicate $comms(O)$ generates a communication constraint for messages to and from the object, which is initially defined as follows.

$$\bigwedge_{p\text{-}O \in O.env} \Box(p \Rightarrow p\text{-}O) \quad \wedge \quad \bigwedge_{q\text{-}O \in O.comp} \Box(q\text{-}O \Rightarrow q)$$

The intuition behind this constraint is that it ensures that when a message, corresponding to predicate p, arrives from the object's environment, not only is p made true, but also the renamed version of p used within the object is also made true. Similarly, when a component predicate is made true in the object, the appropriate message predicate must be broadcast. We will see in §3.3 that this constraint can be refined in order to represent communication delays.

Before providing the semantics of rules within objects, we will remove the first of our constraints. This allows us to introduce a dense temporal logic in order to model execution; it is in this logic that the semantics of individual rules will be presented.

3.2 Asynchronous Execution

The first constraint we remove is that of objects executing in synchrony. This allows asynchronous execution, where each object may progress at its own speed and using its own local clock. Thus, although each individual object sees time (execution) as a sequence of discrete moments (steps), the behaviour of the whole system can be represented using a dense model, where the intervals over which a particular object is active are labelled by that object's name. Consequently, associated with each object, O, is a clock, modelled by a predicate '$tick(O)$', which provides the model structure for that object. For example, in Figure 6.1, an object's state is mapped directly to the appropriate *tick* interval within the dense model provided by the logic (called TLR — see later).

In this way, we can guarantee that each object's states correspond to *tick* intervals, and can ensure that if a predicate is to be satisfied in a particular state of a particular object's execution, the predicate will be satisfied in the interval corresponding to that state.

As objects execute asynchronously, the '*tick*' intervals for various clocks are allowed to overlap. For example, consider four Concurrent METATEM objects, O_1, O_2, O_3, and O_4. The clocks associated with each particular object might overlap as shown in Figure 6.2.

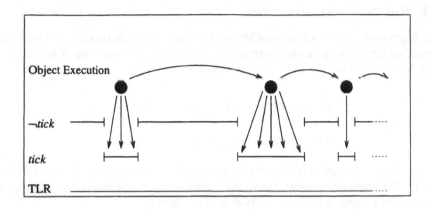

Fig. 6.1. Mapping discrete execution onto a dense model.

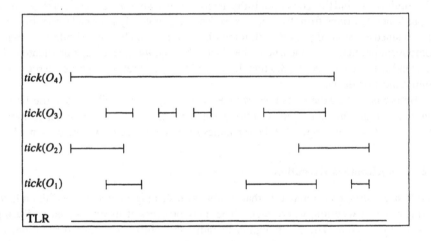

Fig. 6.2. Mapping several discrete executions onto a dense model.

In order to represent this, logically, we use a temporal logic based upon a dense model of time and overlay 'tick' intervals for various objects upon this. The logic we use is called the 'Temporal Logic of the Reals' (TLR), which was originally developed to enable the *abstract* specification of reactive systems [5]. In dense temporal logics, there is no concept of an unique *next* state — between every two states there are an infinite number of other states. As the name suggests, TLR is based upon the domain consisting of the positive real numbers, while the logic described in §2.2 is based upon the domain of natural numbers. Though a full definition of TLR can be found in [5], we give a brief outline of the logic below.

TLR: As, in TLR, the model of time is isomorphic to the the real numbers, rather than the integers, there is no distinguished *next* moment in time. Thus, the only temporal operators within TLR refer to intervals[3]:

$A\,\mathcal{U}^+ B$ — meaning A is true at any moment in the interval between now and the moment when B becomes true.

$C\,\mathcal{S}^- D$ — meaning C was true at any moment in the interval between the moment when D was true and now.

These are the two basic operators within the logic. The only other constraint added in the original TLR definition [5] that we make use of here is that of *finite variability*. The definition of finite variability states that between any two points, each predicate can only vary its value a finite number of times. This avoids the case where predicates 'toggle' values infinitely within a finite period.

As described above, we will define a new predicate '*tick*' within TLR as follows[4].

$$(\textbf{start} \;\Rightarrow\; tick) \;\wedge\; \Box(\neg tick \;\Rightarrow\; (\neg tick)\,\mathcal{U}^+ tick) \;\wedge\; \Box(tick \;\Rightarrow\; tick\,\mathcal{U}^+(\neg tick))$$

Thus, '*tick* periods' occur infinitely often, separated by '¬*tick* periods'. Note that each *tick* and ¬*tick* interval is arbitrary, but finite, as each such interval is followed *sometime in the future* by a period in which the opposite *tick* occurs. Note that, this infinite occurrence, together with the *finite variability* condition, ensures that all the domain is 'covered' by periods of *tick*/¬*tick* (i.e., an infinite number of alternations between *tick*/¬*tick* is not allowed in any finite period).

We extend this definition so that '*tick*' is parameterised by the particular object to which it refers (as shown in Figure 6.2). This represents the fact that individual objects have their own clock. We will model a predicate, p, being satisfied in a state, s, by the predicate p-*Obj* being satisfied in the *tick* associated with s.

Definition of $\mathcal{M}_{rules}$ Earlier, we gave the semantics of an individual object in terms of the semantics of the rules within that object. Now we can give the semantics of individual rules as a TLR formula. This formula is then applied within each interval where the appropriate *tick* occurs. To do this, we require auxiliary predicates next-tick, last-tick and during-tick, which can be described as follows.

next-tick(O, X) — is satisfied if formula 'X' is satisfied within the next 'O' tick.
last-tick(O, X) — is satisfied if formula 'X' is satisfied within the last 'O' tick
(**false** if there was no last tick).
during-tick(O, X) — is satisfied if formula 'X' is satisfied throughout the current
'O' tick.

Using these predicates, we can define the semantics of an object's set of internal rules as follows.

$$\mathcal{M}_{rules}(O) \;=\; \bigwedge_{R \in\, O.Rules} \text{next-tick}(O, \mathcal{M}^O_{rule}(R)) \;\wedge\; \text{next-tick}(O, \mathcal{M}_{rules}(O))$$

---

[3] Note that we have extended TLR, in the obvious way, to incorporate a finite past.
[4] We assume that both the **start** and '$\Box$' operators, are defined.

Note that $\mathcal{M}_{rules}$ is defined recursively, as we must expand the semantics of each object's set of rules at each state in the object's execution sequence.

Now, as any Concurrent METATEM rule can be transformed into one of the four different types of SNF rule [13], we need only give the semantics for each of these types of rule. These are given below, with C representing a conjunction of (non-temporal) literals, D representing a disjunction of (non-temporal) literals, and l representing a single (non-temporal) literal.

$$\mathcal{M}^O_{rule}(\textbf{start} \Rightarrow D) = [(\neg\text{last-tick}(O, \textbf{true})) \Rightarrow \text{during-tick}(O, D)]$$

$$\mathcal{M}^O_{rule}(\textbf{start} \Rightarrow \Diamond l) = [(\neg\text{last-tick}(O, \textbf{true})) \Rightarrow \Diamond(tick(O) \wedge \text{during-tick}(O, l))]$$

$$\mathcal{M}^O_{rule}(\bullet C \Rightarrow D) = [\text{last-tick}(O, C) \Rightarrow \text{during-tick}(O, D)]$$

$$\mathcal{M}^O_{rule}(\bullet C \Rightarrow \Diamond l) = [\text{last-tick}(O, C) \Rightarrow \Diamond(tick(O) \wedge \text{during-tick}(O, l))]$$

Note that, in order to ensure that predicates are checked at the appropriate moment, we add the constraint that if a predicate becomes true within a *tick* period, then it remains true until the end of that periond, i.e.,

$$(p \wedge tick(O)) \Rightarrow p\,\mathcal{U}^+\neg tick(O)$$

This ensures that, if we need to check whether a certain predicate was true during the last period, we need only check the predicate's value at the end of the period.

Synchronous Semantics Before continuing removing restrictions from the basic Concurrent METATEM system that we are modelling, we note how the synchronous version can be represented in TLR. This is relatively easy, once we have the semantics of the asynchronous version in that all we must do is to 'link' all the local clocks together. Thus, by ensuring that all the local clocks are exactly the same, we have effectively introduced a global clock enforcing synchronous execution. This linkage can simply be carried out by adding the following constraint.

$$\forall i.\ \forall j.\ \Box(tick(O_i) \Leftrightarrow tick(O_j))$$

Here, a '*tick*' within one object is exactly the same as a '*tick*' within any other.

Although we can define the semantics of synchronous execution in this way, it would obviously be preferable, if synchrony were required, to use a discrete model in the first place.

3.3 Delaying Messages

We now remove the constraint that messages are instantaneous. This provides a more natural computational model whereby messages sent are only guaranteed to arrive at *some* time in the future. Further, since we now have asynchronous execution it makes little sense to try to ensure that messages sent are received by all objects at the same time. Thus, to enforce this new model of communication, we simply replace the above definition of *comms(O)* by

$$\bigwedge_{p\text{-}O\in O.env} \Box(p \Rightarrow \Diamond p\text{-}O) \quad \land \quad \bigwedge_{q\text{-}O\in O.comp} \Box(q\text{-}O \Rightarrow \Diamond q)$$

Again, *p-O* and *q-O* range over the sets of environment and component predicates of object *O*, respectively. This constraint ensures that when a message, corresponding to predicate *p*, arrives from the object's environment, a commitment to make true the renamed version of *p* used within the object is given. Similarly, when a component predicate is made true in the object, the appropriate message predicate must eventually appear in the environment.

The only minor problem with this new definition is what to do about messages that arriving at an object while it is not executing, i.e., when *tick* is false? In order to cope with this, the above definition can be changed to

$$\bigwedge_{p\text{-}O\in O.env} \Box(p \Rightarrow \Diamond(tick(O) \land p\text{-}O)) \quad \land \quad \bigwedge_{q\text{-}O\in O.comp} \Box(q\text{-}O \Rightarrow \Diamond q)$$

thus ensuring that the received message is recorded during a *tick* of the object.

3.4 Synchronisation

If we now allow environment predicates in the right-hand sides of object rules, we can achieve synchronisation. The side-effect of synchronising objects occurs naturally so long as we ensure that objects can not define the truth value of environment predicates. Under this assumption, then if an object is waiting for an environment predicate to be satisfied, it must wait for the appropriate message to be received. This occurs naturally in the model as the current *tick/¬tick* interval can be 'stretched' out for a finite amount of time while waiting for such a message.

In order to ensure that objects can not directly affect the truth value of an environment predicate, we simply add a constraint that says that such a predicate must be false if no new message has been received since the last message of this type was recorded (or if no such message has *ever* been received), i.e.[5]

$$[(\neg p) \, \mathcal{S}^-(tick(O) \land p\text{-}O) \, \lor \, \blacksquare \neg p\text{-}O] \quad \Rightarrow \quad \neg p\text{-}O$$

3.5 Structuring the Object Space

By introducing the group structuring of objects, we effectively change the communication constraint so that messages are only broadcast to members of particular groups. To give an idea of how this can be reflected in the semantics, assume that a group is simply represented as a structure containing the names of the objects within the group,

---

[5] Here, '$\blacksquare$', representing "always in the past", can be defined using '$\mathcal{S}^-$'.

and that there are extra *Group* and *Name* fields within the object structure. Now, consider the following communication constraint in which it is assumed that each object is a member of one group only.

$$\bigwedge_{p\text{-}O \in O.env} \Box((p(G) \land O.Name \in G) \Rightarrow \Diamond(tick(O) \land p\text{-}O))$$

$$\land$$

$$\bigwedge_{q\text{-}O \in O.comp} \Box(q\text{-}O \Rightarrow \Diamond q(O.Group))$$

Thus, a predicate broadcast includes an extra argument representing the group of the sender. Any receiver can only record such a message if it is also a member of that group.

3.6 Dynamic Interfaces

A more complex situation occurs when we wish to model the ability of an object to dynamically change its interface, for example using listen or ignore. Here, since execution of a rule might change the syntactic representation of the object itself, for example the *Env* or *Comp* fields, semantic functions such as $\mathcal{M}^O_{rule}$ must be redefined to return not only a TLR formula, but the updated version of the object structure that will be used for the semantics of the next execution step.

Thus, we have outlined the basic approach to the construction of a logical semantics for Concurrent METATEM. This semantics can be expanded and extended still further in order to incorporate more complex mechanisms, such as those for ensuring that predicates have particular default values (e.g., default to **false**) and for modelling the manipulation of the message queue by an object.

4   Conclusions and Future Work

While Concurrent METATEM provides a novel model for the simulation and implementation of reactive systems, its semantic clarity is slightly compromised by the operational model used. In particular, the fact that objects can execute asynchronously means that a straight-forward semantics given using a discrete temporal logic is inappropriate. However, we have shown that by utilising a dense temporal logic and overlaying this with the execution sequence for each individual object, we are able to model a range of properties of asynchronous Concurrent METATEM

While we have provided the basis for a semantics, this paper still only represents a step towards a full semantics for Concurrent METATEM. Future work will cover the more esoteric features of Concurrent METATEM (some of which have yet to be implemented), such as dynamic object creation and meta-level manipulation. Similarly, a full semantics would take account of the backtracking and commit that can occur within object execution. Further in the future, we wish to look at alternatives to temporal semantics for Concurrent METATEM. For example, we are considering using an algebraic framework which incorporates the notion of broadcast message-passing [23].

Finally, we note that TLR has been utilised in a similar way in order to provide a semantics for the MYWORLD DAI testbed [26].

References

1. G. Agha. *Actors - A Model for Concurrent Computation in Distributed Systems*. MIT Press, 1986.

2. G. Andrews. Paradigms for Process Interaction in Distributed Programs. *ACM Computing Surveys*, 23(1):49–90, March 1991.

3. H. Bal, J. Steiner, and A. Tanenbaum. Programming Languages for Distributed Computing Systems. *ACM Computing Surveys*, 21(3):261–322, September 1989.

4. H. Barringer, M. Fisher, D. Gabbay, G. Gough, and R. Owens. METATEM: A Framework for Programming in Temporal Logic. In *Proceedings of REX Workshop on Stepwise Refinement of Distributed Systems: Models, Formalisms, Correctness*, Mook, Netherlands, June 1989. (Published in *Lecture Notes in Computer Science*, volume 430, Springer Verlag).

5. H. Barringer, R. Kuiper, and A. Pnueli. A Really Abstract Concurrent Model and its Temporal Logic. In *Proceedings of the Thirteenth ACM Symposium on the Principles of Programming Languages*, St. Petersberg Beach, Florida, January 1986.

6. K. Birman and T. Joseph. Reliable Communication in the Presence of Failures. *ACM Transactions on Computer Systems*, 5(1):47–76, February 1987.

7. K. Birman. The Process Group Approach to Reliable Distributed Computing. Techanical Report TR91-1216, Department of Computer Science, Cornell University, July 1991.

8. A. Borg, J. Baumbach, and S. Glazer. A Message System Supporting Fault Tolerance. In *Proceedings of the Ninth ACM Symposium on Operating System Principles*, New Hampshire, October 1983. ACM. (In ACM Operating Systems Review, vol. 17, no. 5).

9. E. A. Emerson. Temporal and Modal Logic. In J. van Leeuwen, editor, *Handbook of Theoretical Computer Science*. Elsevier, 1990.

10. M. Fisher. A Survey of Concurrent METATEM — The Language and its Applications. In *First International Conference on Temporal Logic (ICTL)*, July 1994.

11. M. Fisher and R. Owens. From the Past to the Future: Executing Temporal Logic Programs. In *Proceedings of Logic Programming and Automated Reasoning*, St. Petersberg, Russia, July 1992. (Published in *Lecture Notes in Computer Science*, volume 624, Springer-Verlag).

12. M. Fisher and M. Wooldridge. A Logical Approach to the Representation of Societies of Agents. In *Proceedings of Second International Workshop on Simulating Societies (SimSoc)*, Certosa di Pontignano, Siena, Italy, July 1993.

13. M. Fisher. A Normal Form for First-Order Temporal Formulae. In *Proceedings of Eleventh International Conference on Automated Deduction (CADE)*, Saratoga Springs, New York, June 1992. (Published in *Lecture Notes in Computer Science*, volume 607, Springer-Verlag).

14. M. Fisher. Concurrent METATEM — A Language for Modeling Reactive Systems. In *Parallel Architectures and Languages, Europe (PARLE)*, Munich, Germany, June 1993. Springer-Verlag.

15. D. Gabbay. Declarative Past and Imperative Future: Executable Temporal Logic for Interactive Systems. In B. Banieqbal, H. Barringer, and A. Pnueli, editors, *Proceedings of Col-*

loquium on Temporal Logic in Specification, Altrincham, U.K., 1987. (Published in *Lecture Notes in Computer Science*, volume 398, Springer-Verlag).

16. D. Harel and A. Pnueli. On the Development of Reactive Systems. Technical Report CS85-02, Department of Applied Mathematics, The Weizmann Institute of Science, Revohot, Israel, January 1985.

17. C. Hewitt. Control Structure as Patterns of Passing Messages. In P. H. Winston and R. H. Brown, editors, *Artificial Intelligence: An MIT Perspective (Volume 2)*. MIT Press, 1979.

18. K. Kahn, E. Tribble, M. Miller, and D. Bobrow. Vulcan: Logical Concurrent Objects. In B. Shriver and P. Wegner, editors, *Research Directions in Object-Oriented Programming*. MIT Press, 1987.

19. H. Lieberman. Concurrent Object-Oriented Programming in Act 1. In A. Yonezawa and M. Tokoro, editors, *Object-Oriented Concurrent Programming*. MIT Press, 1986.

20. Z. Manna and A. Pnueli. *The Temporal Logic of Reactive and Concurrent Systems: Specification*. Springer-Verlag, New York, 1992.

21. T. Maruichi, M. Ichikawa, and M. Tokoro. Modelling Autonomous Agents and their Groups. In Y. Demazeau and J. P. Muller, editors, *Decentralized AI 2 – Proceedings of the 2nd European Workshop on Modelling Autonomous Agents and Multi-Agent Worlds (MAAMAW '90)*. Elsevier/North Holland, 1991.

22. A. Pnueli. Applications of Temporal Logic to the Specification and Verification of Reactive Systems: A Survey of Current Trends. *Lecture Notes in Computer Science*, 224, August 1986.

23. K. V. S. Prasad. A Calculus of Broadcasting Systems. In S. Abramsky and T. Maibaum, editors, *Proceedings of the International Joint Conference on Theory and Practice of Software Development (TAPSOFT)*, Brighton, U.K., April 1991. (Published in *Lecture Notes in Computer Science*, volume 493, Springer Verlag).

24. E. Shapiro and A. Takeuchi. Object Oriented Programming in Concurrent Prolog. In E. Shapiro, editor, *Concurrent Prolog–Collected Papers*, chapter 29. MIT Press, 1987.

25. R. Strom and S. Yemini. Optimistic Recovery in Distributed Systems. *ACM Transactions on Computer Systems*, 3(3):204–226, August 1985.

26. M. Wooldridge. This is MYWORLD: The Logic of an Agent-Oriented Testbed for DAI. In M. Wooldridge and N. R. Jennings, editors, *Intelligent Agents — Proceedings of the 1994 Workshop on Agent Theories, Architectures, and Languages*, Springer-Verlag, 1995.

27. A. Yonezawa, editor. *ABCL: An Object-Oriented Concurrent System*. MIT Press, 1990.

28. A. Yonezawa and M. Tokoro. Object-Oriented Concurrent Programming. In A. Yonezawa and M. Tokoro, editors, *Object-Oriented Concurrent Programming*. MIT Press, 1986.

Constraint Deduction in an Interval-based Temporal Logic

Jana Koehler & Ralf Treinen

Abstract. We describe reasoning methods for the interval-based modal temporal logic LLP which employs the modal operators *sometimes*, *always*, *next*, and *chop*. We propose a constraint deduction approach and compare it with a sequent calculus, developed as the basic machinery for the deductive planning system PHI which uses LLP as underlying formalism.

1 Motivation

The work presented in this paper was motivated by an application coming from the field of deductive planning. In the PHI system [BBD+93] planning is done on the formal basis of an interval-based modal temporal logic. Apart from plan generation, the reuse and the modification of existing plans is also investigated. Since plan generation and plan reuse are formalized as deductive processes, various proofs in the underlying temporal logic have to be performed raising the need for an efficient proof method for the logic.

In deductive planning, plans are generated by *constructively* proving plan specifications, for example formulae of the form

$$pre \wedge \mathsf{Plan} \to goals$$

which describe the properties of the desired plan: if Plan is carried out in a situation where the preconditions *pre* hold then the *goals* will be reached.

During the proof, the plan metavariable Plan is replaced by a plan (formula) that satisfies the specification. Plan formulae in LLP [BDK92] provide constructs that allow, e.g. to

- sequentially compose plans from arbitrary subplans or atomic actions,
- incorporate control structures for conditional and iterative plans, and
- abstract from the exact execution time of actions in plans.

When plans are executed, every action leads to a new state of the world, i.e. plans describe temporally ordered sequences of states. This suggests using a modal temporal logic as the underlying formalism for deductive planning and grounding its semantics on *intervals* in contrast to the usual *possible worlds* semantics. Intervals can be seen as possible worlds to which an additional structure has been added, i.e. by considering worlds as sequences of states. This sequential structure of the worlds reflects the semantics of plans.

The interval-based modal temporal logic LLP [BDK92] is consequently developed as a formal basis combining features of *choppy logic* [RP86] with a *temporal logic for programs* [Krö87].

The paper is organized as follows: in Section 2, we describe related work. We review the logic LLP in greater detail and discuss the executability of LLP plans in Section 3. Section 4 describes a sequent calculus approach for deductive planning and plan reuse and discusses its advantages in the underlying context. As an alternative to the LLP sequent calculus we introduce a constraint deduction approach in Section 5. Finally, we conclude with some discussion comparing both approaches in Section 6 and show how the constraint deduction approach is applied for tasks of temporal abstraction.

2   Related Work

Interval-based temporal logics have been proposed by several authors [MM83, Gab89] as appropriate formalisms to describe the behavior of programs or plans. Plans can be decomposed into successively smaller periods or intervals of, e.g. subplans or actions. The intervals provide a convenient framework for introducing quantitative timing details. State transitions can be characterized by properties relating the initial and final values of variables over intervals of time [MM83].

The logic LLP which is considered in this paper is a first order extension of the propositional linear temporal logic PTL(U,X,C) [RP86]. PTL(U,X,C) contains the modal operators *weak-next, until* and *chop* and has an interval-based semantics. The concept of *local variables*, the interpretation of which may vary from state to state, was borrowed from [Krö87, MM83, Hal87] in order to describe the action to be performed in a state as well as the effects of the action.

In [Gab89, FO92] a *declarative* as well as an *imperative* view on temporal logic formulae is proposed:

> "A future statement of temporal logic can be understood in two ways: the declarative way, that of describing the future as a temporal extension; and the imperative way, that of making sure that the future will happen the way we want it." (cf. [Gab89],page 402)

Grounded on this view, the logic USF has been developed in [Gab89]. Specification formulae in USF can be automatically re-written into an executable form utilizing an *exec* predicate. For example, *exec*(a_1) will make a_1 true. The re-written formula is an *equivalent* logical formulation of the specification.

Plan specification formulae in LLP can also be viewed as declaratively describing the future as a temporal extension. To obtain a plan, specification formulae are not equivalently transformed but are constructively proved, i.e. an example (plan) is constructed which *satisfies* the specification. The plan can be seen as a program for controlling process behavior: its execution in the initial state is sufficient to reach the specified goals. This view led to the *plans are programs* paradigm which has already been proposed by, e.g. [Bib86, MW87].

In order to benefit from the representational advantages provided by modal logics, reasoning mechanisms for modal formulae have to be developed. Our work fits into the framework of translation oriented methods similar to those described in [Ohl91, FS91]. It extends the constraint deduction approach for serial modal logics with *sometimes* and *always* [FS91] to a non-serial modal temporal logic with the additional modal operators *next* and *chop*.

3   The Interval-based Modal Temporal Logic LLP

LLP (Logical Language for Planning) [BDK92] is an interval-based modal temporal logic which combines features of *choppy logic* [RP86] with a *temporal logic for programs* [Krö87]. The basis of LLP is a many-sorted first order language. Furthermore, we distinguish *local variables*, the value of which may vary from state to state and *global* variables which are the usual logical variables. Local variables are borrowed from *programming logics* where they correspond to program variables.

LLP provides the modal operators $\circ$ *(next)*, $\Diamond$ *(sometimes)*, $\Box$ *(always)*, and the binary modal operator ; *(chop)*. Furthermore, control structures like *if-then-else* and *while* are available. In the following we shortly review the main properties of LLP as introduced in [BDK92].

A state σ_i is a valuation assigning domain elements to local variables. Note that only the values of local variables may change from state to state. Function and predicate symbols are *rigid*, i.e. their interpretation does not vary over time. An interval σ is a nonempty finite or infinite sequence of states $\langle \sigma_0 \sigma_1 \ldots \rangle$. W denotes the set of all intervals. The length of an interval σ is defined as

$$|\sigma| = \begin{cases} \omega, & \text{if } \sigma \text{ is infinite} \\ n, & \text{if } \langle \sigma = s_0 s_1 \ldots s_n \rangle \end{cases}$$

Observe that $|\sigma| = 0$ iff $\sigma = \langle \sigma_0 \rangle$ is a *singleton* containing only one state. Intuitively, the length of an interval does not represent the number of states this interval contains, but the number of possible state transitions. The *immediate accessibility* on intervals is defined as the subinterval relationship R with

$$\sigma R \sigma' \quad \text{iff} \quad \sigma = \langle \sigma_0 \sigma_1 \sigma_2 \ldots \rangle \text{ and } \sigma' = \langle \sigma_1 \sigma_2 \ldots \rangle.$$

R^* denotes the transitive and reflexive closure of R and R^+ denotes the transitive closure of R. The *composition* is defined as a partial function over the set of intervals W:

$$\sigma \circ \sigma' = \begin{cases} \sigma, & \text{if } \sigma \text{ is infinite} \\ \langle \sigma_0 \ldots \sigma_{n-1} \sigma_n \sigma_{n+1} \ldots \rangle, & \text{if } \begin{array}{l} \sigma = \langle \sigma_0 \ldots \sigma_{n-1} \sigma_n \rangle \text{ and} \\ \sigma' = \langle \sigma_n \sigma_{n+1} \ldots \rangle \end{array} \end{cases}$$

Global variables are interpreted by mapping them to domain elements using a valuation function. The value of a local variable in an interval σ for a given interpretation is its value in the initial state of the interval. The satisfiability relation $\models$ for modal-free formulae is defined as in classical first order logic. F and T denote the propositional constants *false* and *true*, respectively. For the modal operators we define:

- $\sigma \models_I \circ\phi$ iff $\sigma' \models_I \phi$ for all $\sigma' \in W$ with $\sigma R \sigma'$
- $\sigma \models_I \Diamond\phi$ iff $\sigma' \models_I \phi$ for some $\sigma' \in W$ with $\sigma R^* \sigma'$
- $\sigma \models_I \Box\phi$ iff $\sigma' \models_I \phi$ for all $\sigma' \in W$ with $\sigma R^* \sigma'$
- $\sigma \models_I \phi\,;\psi$ iff there are $\sigma', \sigma'' \in W$, with $\sigma = \sigma' \circ \sigma''$, σ' finite and $\sigma' \models_I \phi$ and $\sigma'' \models_I \psi$

The immediate accessibility relation R is not serial, i.e.

$$\forall\sigma \, \exists\sigma' \; \sigma R \sigma'$$

does not hold since an interval of length zero has no successor. For example, $\circ F$ holds in an interval σ iff σ has length 0, i.e. it is a singleton. More generally, $\circ^n F$ holds in σ iff σ has at most n states, that is iff σ has at most length $n-1$. A formula $\phi \wedge \neg\circ F \wedge \circ\circ F; \circ\Box\psi$ holds in an interval $\langle \sigma_0\sigma_1\sigma_2\sigma_3 \dots \rangle$ if

- ϕ holds in the subinterval $\langle \sigma_0\sigma_1 \rangle$ and
- $\circ\Box\psi$ holds in the subinterval $\langle \sigma_1\sigma_2\sigma_3 \dots \rangle$, i.e., ψ holds in all subintervals $\langle \sigma_n \dots \rangle$ with $n \geq 2$.

3.1 Properties of LLP

Several results from the literature help to characterize LLP with respect to the expressive power of the modal operators and their axiomatization. The modal operators $\Box$ and $\Diamond$ can be expressed by ; [RP86] using the axioms $\Diamond\phi \leftrightarrow T;\phi$ and $\Box\phi \leftrightarrow \neg\Diamond\neg\phi$.

Further results concern the axiomatization of first order temporal logics. An axiomatization of the propositional linear temporal logic PTL(U,X,C) is developed and a complete and sound decision procedure based on semantic tableaux is given for the logic in [RP86].

Szalas proved in [Sza86] that there is no finistic and complete axiomatization of first order temporal logics of linear and discrete time and gives an infinitary complete proof system for the logic in [Sza87]. In [SH88] it is proven that a first order temporal logic with *equality* and *until* is both weakly and strongly incomplete. A logic is defined to be *weakly incomplete* if "the set of all tautologies (over an arbitrary signature) of the logic is not recursively enumerable or, equivalently, if there is no finistic proof system which is sound and complete for the logic". A logic is *strongly incomplete* if "for no signature the set of tautologies over this signature is recursively enumerable or, equivalently, if the set of tautologies over the empty signature is not recursively enumerable".

The results of Szalas hold for LLP as well with the consequence that there is no sound and complete first order calculus for LLP.

3.2 Planning in LLP

Plan generation is carried out by *constructively* proving plan specification formulae in a sequent calculus [BDK92]. One type of plan specifications used by the PHI system, is *liveness properties* containing in their goal specification a temporally ordered sequence of intermediate subgoal states, e.g. formulae of the form:

$$pre \wedge \text{Plan} \rightarrow \Diamond(goal_1 \wedge goal_2 \wedge \Diamond(goal_3 \wedge \Diamond(goal_4)))$$

As a result of the proof, a plan formula is obtained which is sufficient for the plan specification. The plan utilizes the local variable *ex*, the value of which is a term representing the action to be executed in the current state. An example of a plan formula reads

$$ex = action_1 \wedge \neg \circ F \wedge \circ \circ F ; \Box \phi ; ex = action_2 \wedge \neg \circ F \wedge \circ \circ F ; ex = action_3$$

describing a plan which contains three actions. The subplan containing the actions $action_2$ and $action_3$ can be executed at an arbitrary time after $action_1$ has been executed provided that a formula ϕ always holds between the execution of $action_1$ and $action_2$. ϕ describes in our context the minimal preconditions for the subsequent actions which must be adhered to. This plan formula holds in an interval $\sigma = \langle \sigma_0 \sigma_1 ... \rangle$ if

- $ex = action_1$ holds in the first state σ_0 of σ,
- there is an $n \geq 1$ such that all intervals $\langle \sigma_1 ... \sigma_n \rangle$ to $\langle \sigma_n \rangle$ satisfy ϕ,
- $ex = action_2$ holds in σ_n, and
- $ex = action_2$ holds in σ_{n+1}.

3.3 Executability of LLP Plans

The PHI planner is able to generate complex plans containing control structures like iteration and case analysis. One prototypical application domain of PHI is a subset of UNIX, namely the mail domain. Here, the planner generates *abstract* plans that are used by a plan recognizer to identify the goals of a user and to offer active help. The following formula shows such an example plan achieving the goal *"Read all mail from sender Joe"*:

```
n := 1;
while n < length(system_mbox) do
        if sender(msg(n, system_mbox)) = joe
            then ex = type(n, system_mbox)
            else ex = empty_action;
n := n + 1
od
```

In order to execute an abstract plan, a *plan interpreter* [Den94] is used, which performs sensing actions in the application system in order to instantiate variables and parameters occurring in the abstract plan. Furthermore, since control structures cannot be executed directly in the application system they are replaced by executable action sequences according to their definition. The plan starts by instantiating the internal counter n with value 1, which refers to the current mail in the mailbox. The while–loop is stepwisely expanded according to the definition of the while-operator:

$$\textbf{while } \varepsilon \textbf{ do } \alpha \textbf{ od } ; \beta \leftrightarrow \textbf{if } \varepsilon \textbf{ then } \alpha ; [\textbf{while } \varepsilon \textbf{ do } \alpha \textbf{ od } ; \beta] \textbf{ else } \beta$$

The resulting case analysis is resolved by testing the validity of the condition $n <$ $length(system\_mbox)$ in the application system. According to the definition of the if-operator, the non-valid branch is eliminated:

$$\textbf{if } \varepsilon \textbf{ then } \alpha \textbf{ else } \beta \leftrightarrow [\varepsilon \to \alpha] \wedge [\neg\varepsilon \to \beta]$$

Let us assume that the mailbox $system\_mbox$ of the user contains 3 mails with the first and last from sender Joe. In this situation, the following executable action sequence is obtained as a refinement of the abstract plan:

$ex = type(1, system\_mbox);$	read the first mail
$ex = empty\_action;$	skip the second mail
$ex = type(3, system\_mbox)$	read the last mail

4   The Sequent Calculus Approach

Together with the logic LLP a sequent calculus was developed which is used by the planner for the constructive proofs of specifications. The sequent calculus extends the S4 sequent calculus (see for example [Wal89]) with rules for the additional modal operators o and ; and with derived rules which are of importance for deductive planning and plan reuse.

Typical examples of such rules are the *right*-o rule, the *chop composition* and the *sometimes-to-next* rule [BDK92].

- *right*-o: $\dfrac{\Gamma^* \Rightarrow A, \Delta^*}{\Gamma \Rightarrow \circ A, \Delta}$ with $\begin{aligned}\Gamma^* &= \{B | \circ B \in \Gamma\} \cup \{\Box B | \Box B \in \Gamma\}, \text{ and}\\ \Delta^* &= \{B | \circ B \in \Delta\} \cup \{\Diamond B | \Diamond B \in \Delta\}\end{aligned}$

- *chop composition*: $\dfrac{\phi_1 \Rightarrow \psi_1 \qquad \phi_2 \Rightarrow \psi_2}{\phi_1 ; \phi_2 \Rightarrow \psi_1 ; \psi_2}$

- *sometimes-to-next*: $\dfrac{\Gamma \Rightarrow \circ\phi \wedge \neg\circ F, \Delta}{\Gamma \Rightarrow \Diamond\phi, \Delta}$

To guide sequent rule applications during a proof, a tactic language is provided in which proof tactics can be described [Den94]. Proof tactics implement search strategies that restrict the search space in the proof and thereby help to reduce unnecessary search effort.

Furthermore, proof tactics specify a certain ordering of modal rule applications. As discussed in [Wal89], the ordering in which modal rules are applied during a sequent derivation influences whether a proof is obtained or not. If an inappropriate order is chosen, a proof may not be found. The cause of this order dependence is the fact that some modal rules lead to formulae being "deleted", i.e. subformulae from the conclusion of a sequent do no appear in the premise, see for example the *right*-o rule. Therefore, besides restricting the search space, proof tactics help to maintain sufficient formulae in the sequents to complete the proof.

The use of tactics to guide proofs in the LLP calculus led to an efficient implementation of plan generation and plan reuse in the PHI system [BBD+93]. The tactic language which is provided by the system supports the formulation of new tactics when new proof tasks must be solved by the system or when the syntactical class of formulae, on which these proofs are performed, is extended.

Practical experiences, e.g. comparing the effort for plan modification to plan generation [Koe94b] show that proof tactics lead to a very efficient reasoning in the underlying logic. Hence, proof tactics can be seen as one way of developing efficient reasoning methods for modal logics based on Gentzen type calculi.

On the other hand, such classical calculi for modal logics and its extensions, e.g. temporal logic and dynamic logic, which are usually of Hilbert, Gentzen or Tableaux type are often criticized for their inefficiency because "the branching rate in the search space is very high" and because they " require special implementations of deduction systems" (cf. [Ohl91]).

Alternatively, a translation of modal formulae into predicate logic syntax such that standard predicate logic deduction systems are applicable has been proposed by some authors [Ohl91, FS91] for modal logics with possible worlds semantics which provide the modal operators $\Diamond$ and $\Box$.

In the work presented in [Ohl91] a *functional* and a *relational* translation method are developed. The target calculus for the formulae obtained by the translation is the resolution calculus. The relational translation introduces special predicates representing the accessibility relations. This approach is very flexible because different kinds of accessibility relations can easily be handled but the number and the size of translated clauses is increased by the literals which are necessary to represent the possible worlds. Since standard resolution strategies do not differentiate between "normal" and "special" predicates, many unnecessary resolution steps may occur.

To overcome the problem, the functional translation method has been developed whereby the relevant information about worlds is represented in terms, and reasoning about possible worlds is done with specialized unification algorithms. On one hand, this method leads to a more efficient calculus, while on the other hand, it is not known whether this method can be extended to more expressive modal logics, which, for example, contain the *chop* operator.

A generalization of the relational translation method, restricted to modal logics with serial accessibility relations, is presented in [FS91]. A translation into a first order constraint logic is proposed and a hybrid reasoning method combining ordinary deduction with special purpose methods for constraint processing is developed.

During the development of the sequent calculus approach for LLP the question arose whether these translation oriented methods can be extended in order to provide a framework for a semantics-based translation of LLP.

Furthermore, when such a translation of LLP into first order logic is possible, the question arises as to how an efficient calculus can be developed which might serve as an alternative to the sequent calculus approach. It turns out that this remains a difficult problem even for the translated logic.

5   The Constraint Deduction Approach

In the previous section we described the motivation that led to the development of a semantics-based translation method for LLP. Since the logic we consider is more expressive than those logics for which translation oriented methods already exist [FS91, Ohl91] we hope to gain further insight into how far these results generalize.

A relational translation is developed during which special predicate symbols are introduced representing the accessibility relations on intervals. Following the method of [FS91], we proceed in three steps in order to obtain a constraint theory and a set of constrained clauses. These can be proven, for instance, by constrained resolution [Bür91], which can be realized by a resolution prover and a satisfiability checker for constraints. The steps are

1. translation of the modal logic LLP into a constraint first order logic CPL by reification of the intervals,
2. transformation of CPL formulae into constrained prenex normal form and
3. skolemization of formulae in constrained prenex normal form.

5.1   The Translation into CPL

The translation of LLP formulae into the constraint predicate logic CPL transforms the modal-logic features of LLP into predicate logic by reification of the intervals. For this purpose we first transform the signature Σ_M of LLP into the signature Σ_P of CPL as follows:

1. CPL contains the two sorts D, denoting the domain of LLP, and W, which denotes the set of intervals.
2. The domain variables in CPL are the global variables from LLP.
3. The function and predicate symbols from LLP carry over to CPL. Observe that, in contrast to [FS91], functions and predicates have a fixed interpretation in all intervals and therefore do not have to be equipped with an extra argument for the actual interval.
4. The local variables of LLP, which may change the interpretation from interval to interval, are translated into unary function symbols of type $(W \rightarrow D)$.
5. For the translation of modal operators we need the additional predicate symbols S (type W), $\geq$ (type WW), $\gg$ (type WW) and $\oplus$ (type WWW). Atoms constructed with these predicates are called *constraints*, we write them in mixfix notation Sx, $x \geq y$, $x \gg y$ and $x = y \oplus z$.

An LLP interpretation $\mathcal{I}_M$ is translated into a CPL interpretation $\mathcal{I}_P$ as follows: The domain of $\mathcal{I}_P$ consists of the domain of $\mathcal{I}_M$ (for sort D) and the set of all intervals (for sort W). The interpretation of global variables and function symbols as well as predicate symbols remains unchanged. The interpretation of a function symbol translating a local variable x is the function that maps an interval σ to the value of x in σ, which is its value in the initial state of σ. The interpretation of the predicate symbol S is the set of all intervals of length 0, $(x, y) \in \gg^{\mathcal{I}_P}$ iff xRy, $(x, y) \in \geq^{\mathcal{I}_P}$ iff xR^*y and $(x, y, z) \in \oplus^{\mathcal{I}_P}$ iff $x = y \circ z$.

As in [Ohl91], we define a translation function π that takes an LLP formula and translates it into a CPL formula. Since intervals are reified in CPL, the actual interval will be explicitly represented in the translated formula. Therefore, π takes as its second argument the variable which refers to the actual interval. The following translation rules from LLP into CPL are used by π:

$$\pi[x, w] := x \quad \text{if } x \text{ a global variable}$$
$$\pi[x, w] := x(w) \quad \text{if } x \text{ a local variable}$$
$$\pi[f(t_1, \ldots, t_n), w] := f(\pi[t_1, w], \ldots, \pi[t_n, w])$$
$$\pi[P(t_1, \ldots, t_n), w] := P(\pi[t_1, w], \ldots, \pi[t_n, w])$$
$$\pi[(F \wedge G), w] := \pi[F, w] \wedge \pi[G, w]$$
$$\pi[(F \vee G), w] := \pi[F, w] \vee \pi[G, w]$$
$$\pi[\neg F, w] := \neg \pi[F, w]$$
$$\pi[\exists x F, w] := \exists x \, \pi[F, w]$$
$$\pi[\forall x F, w] := \forall x \, \pi[F, w]$$
$$\pi[\Diamond F, w] := \exists v(w \geq v \wedge \pi[F, v])$$
$$\pi[\Box F, w] := \forall v(w \geq v \rightarrow \pi[F, v])$$
$$\pi(\circ F, w] := \forall v(w \gg v \rightarrow \pi[F, v])$$
$$\pi[(F;G), w] := \exists v, u(w = v \oplus u \wedge \pi[F, v] \wedge \pi[G, u])$$

Theorem 1 states the soundness of the translation:

Theorem 1 *For all LLP interpretations $\mathcal{I}_M$, LLP formulae ϕ, W-variables w holds*

$$\mathcal{I}_M \models \phi \iff \mathcal{I}_P \models \forall w \, \pi[\phi, w]$$

where $\mathcal{I}_P$ is the translation of $\mathcal{I}_M$.

Proof sketch: The soundness of the translation can be shown by structural induction over terms and formulae. The base case for global variables is trivial because the assignment of global variables does not differ between $\mathcal{I}_M$ and $\mathcal{I}_P$. Proving the soundness of the translation rules for function and predicate symbols is trivial because they are rigid and thus their interpretation remains unchanged. The correctness of the translation rule for local variables follows from the definition of $\mathcal{I}_P$: the interpretation of the function symbol translating a local variable is the value of that variable in the first state of an interval. The cases for normal (non-modal) connectives are straightforward, their interpretations remains unchanged. To prove the correctness of the translation rules for modal operators the LLP semantics as well as the definition of the semantics of the resulting constraint predicates is exploited.   ∎

As a short example the translation of the formula

$$T;P(v) \rightarrow \Diamond P(v)$$

where v is a local variable is shown below. This formula is valid; it is in fact a consequence of the following theorem of LLP:

$$T;\phi \leftrightarrow \Diamond \phi$$

After expanding $\rightarrow$, we obtain ($i_1, i_2, \ldots$ are W-variables):

$$\pi[\neg\Big(T;P(v)\Big) \vee \Diamond P(v), i_1]$$

$$= \neg\pi[T;P(v), i_1] \vee \pi[\Diamond P(v), i_1]$$

$$= \neg\exists i_2, i_3 \Big(i_1 = i_2 \oplus i_3 \wedge \pi[T, i_2] \wedge \pi[P(v), i_3]\Big) \vee \pi[\Diamond P(v), i_1]$$

$$= \neg\exists i_2, i_3 \Big(i_1 = i_2 \oplus i_3 \wedge \pi[T, i_2] \wedge \pi[P(v), i_3]\Big) \vee \exists i_4 \Big(i_1 \geq i_4 \wedge \pi[P(v), i_4]\Big)$$

$$= \neg\exists i_2, i_3 \Big(i_1 = i_2 \oplus i_3 \wedge T \wedge P(v(i_3))\Big) \vee \exists i_4 \Big(i_1 \geq i_4 \wedge P(v(i_4))\Big)$$

5.2   The Transformation into Prenex Normal Form

The formulae computed by the above translation mechanism have the property that all constraints τ occur only in the form $\forall \bar{x} \tau \rightarrow \psi$ or $\exists \bar{x} \tau \wedge \psi$. The aim is to preserve this property during transformation into prenex normal form as this will allow the translation into constrained clauses in the last step. Rules for transforming into prenex normal form maintaining the above property have been given in [Fri91]:

$$\neg\exists \bar{x}(\tau \wedge \phi) \Rightarrow \forall \bar{x}(\tau \rightarrow \neg\phi)$$
$$\exists \bar{x}(\tau \wedge \phi) \wedge \psi \Rightarrow \exists \bar{x}(\tau \wedge \phi \wedge \psi)$$

provided that ψ does not contain variables of $\bar{x}$.[1] The problem is that the transformation

$$\exists \bar{x}(\tau \wedge \phi) \vee \psi \quad \Rightarrow \quad \exists \bar{x}(\tau \wedge [\phi \vee \psi]) \tag{7.1}$$

needs additional conditions in order to be an equivalence transformation. This rule was proven correct in [Fri91] for $\bar{x}$ consisting of the single variable x, τ containing the variable x only and τ being satisfiable in every model under consideration. In [FS91], binary constraints $K(x, y)$ have been considered where y is bound by the surrounding existential quantifier but x is free. There, the correctness condition was that K, denoting the accessibility relation between worlds, is serial.

Generally speaking, (7.1) is an equivalence transformation in the class $\mathcal{C}$ of models if ψ does not contain x and if

$$\mathcal{C} \models \bar{\forall} \exists \bar{x} \, \tau \tag{7.2}$$

---

[1] Only rules for $\exists$ are considered here; the rules for $\forall$ being dual.

holds where $\tilde{\forall}\gamma$ denotes the universal closure of a formula γ. As the reader easily verifies, condition (7.2) subsumes those of [Fri91] for sorted deduction and of [FS91] for reasoning with a serial accessibility relation.

While looking at the constraints introduced by the translation process, it turns out that rule (7.1) applies to the quantifiers which arise from the operators ; and $\lozenge$, as for all CPL interpretations $\mathcal{I}_P$

$$\mathcal{I}_P \models \forall x \exists y \, x \geq y, \text{ and}$$
$$\mathcal{I}_P \models \forall x \exists y, z \, x = y \oplus z.$$

The situation is different with a formula

$$\exists y \Big(x \gg y \wedge \phi \Big) \vee \psi \tag{7.3}$$

as R is not serial. Two cases have to be considered: if the value of x is a sequence of length 0, then (7.3) is obviously equivalent to ψ. If the value of x has a length of at least 1, then we can replace (7.3) by $\exists y(x \gg y \wedge \lfloor \phi \vee \psi \rfloor)$, or alternatively since $\gg$ is functional, by $\forall y(x \gg y \rightarrow \lfloor \phi \vee \psi \rfloor)$. Note that any occurrence of a formula 7.3 is within the scope of a quantifier for x. The case distinction is done by splitting the quantifier for x by using the rule

$$\exists x \gamma \quad \Rightarrow \quad \exists x \Big(Sx \wedge \gamma \Big) \vee \exists x \Big(\neg Sx \wedge \gamma \Big) \tag{7.4}$$

and accordingly for a universal quantifier. As an example, the following formula is transformed into prenex normal form:

$$\forall i_1 \Big\{ \exists i_2 \Big(i_1 \gg i_2 \wedge P(v(i_2)) \Big) \vee Q(w(i_1)) \Big\}$$

We split the quantifier for i_1 using the dual of rule (7.4):

$$\forall i_1 \Big\{ Si_1 \rightarrow \exists i_2 \Big(i_1 \gg i_2 \wedge P(v(i_2)) \Big) \vee Q(w(i_1)) \Big\}$$
$$\wedge \, \forall i_1 \Big\{ \neg Si_1 \rightarrow \exists i_2 \Big(i_1 \gg i_2 \wedge P(v(i_2)) \Big) \vee Q(w(i_1)) \Big\}$$

The prenex normal form is arrived at after three further transformation steps:

$$\forall i_1 \Big\{ Si_1 \rightarrow Q(w(i_1)) \Big\} \wedge \forall i_1 \Big\{ \neg Si_1 \rightarrow \exists i_2 \Big(i_1 \gg i_2 \wedge (P(v(i_2)) \vee Q(w(i_1))) \Big) \Big\}$$
$$\forall i_1, j_1 \Big\{ Si_1 \wedge \neg Sj_1 \rightarrow Q(w(i_1)) \wedge \exists j_2 \Big(j_1 \gg j_2 \wedge (P(v(j_2)) \vee Q(w(j_1))) \Big) \Big\}$$
$$\forall i_1, j_1 \Big\{ Si_1 \wedge \neg Sj_1 \rightarrow \exists j_2 \Big(j_1 \gg j_2 \wedge Q(w(i_1)) \wedge (P(v(j_2)) \vee Q(w(j_1))) \Big) \Big\}$$

5.3   The Skolemization

The rules for the skolemization of formulae in constrained prenex normal form can be found in [Fri91]. In the example of the last subsection, the binary skolem function symbol sk is associated with the existentially quantified variable i_2. We obtain the constrained formula

$$Q(w(i_1)) \wedge (P(v(sk(i_1, j_1))) \vee Q(w(j_1))) \quad / \quad Si_1 \wedge \neg Sj_1$$

as well as the sentence

$$\forall i_1, j_1 \left(Si_1 \wedge \neg Sj_1 \rightarrow j_1 \gg sk(i_1, j_1) \right)$$

which in this case constitutes the *constraint theory*.

5.4   Solving Constraints

In order to obtain a hybrid reasoning system [Fri91] for CPL formulae a constraint solver is needed which can decide the satisfiability of conjunctions of the constraint atoms in a given constraint theory. The constraint theory employed here unfortunately turns out to be undecidable. This can be shown quite easily using the method of [Tre92]. Furthermore, the undecidability of the satisfiability of the constraints follows from the fact that validity of LLP-formulae is undecidable, while constrained resolution is complete relative to the satisfiability of the constraints [Bür91].

Among the positive results on decidability of related constraint systems we mention the seminal paper by Rabin [Rab69], where a decision procedure for the monadic second order theory of strings is given. As a corollary of [Rab69], the full first order theory of *finite* intervals with the predicates $\geq$ and $\gg$ (but without $\oplus$) is decidable.

5.5   Application of the Translation Method

A translation of temporal-logic formulae into first-order logic by reification of intervals leads to an explicit representation of the temporal information contained in the formulae. This property can be used to explicitly reason about temporal relationships. In Section 3 we showed the following example of a plan specification formula in form of a liveness property:

$$pre \wedge \mathsf{Plan} \rightarrow \Diamond(goal_1 \wedge goal_2 \wedge \Diamond(goal_3 \wedge \Diamond(goal_4)))$$

Translating the goal specification formula into a CPL formula and separating the constraints from the first-order part of the formula leads to

$$\exists i_1, i_2, i_3 \; i_1 \geq i_0 \wedge i_2 \geq i_1 \wedge i_3 \geq i_2 \wedge$$
$$goal_1(i_1) \wedge goal_2(i_1) \wedge goal_3(i_2) \wedge goal_4(i_3)$$

Temporal abstraction of plan specifications can now be implemented by removing constraint formulae and thus relaxing the temporal constraints that are specified for a set of subgoals. In the above mentioned example, we require the plan to achieve $goal_1$ and $goal_2$ in the same interval, while $goal_3$ and $goal_4$ have to be achieved later. When we remove the set of constraints, no ordering of subgoal states is required anymore. Temporal abstraction grounded on a manipulation of temporal constraints is used in the PHI system to index plans in a plan library [Koe94a]. This method leads to a well-defined temporal abstraction process and allows to prove a *monotonicity property* that holds between original and abstracted plan specifications. It states that an existing subset relationship between the set of models satisfying two LLP plan specifications is preserved as a subset relationship between the set of models satisfying the abstracted CPL formulae. This property ensures that a retrieval method can be developed that finds a plan solving a given plan specification whenever it exists in the plan library.

6 Conclusion

Two calculi for an interval-based modal temporal logic are discussed in this paper: a sequent calculus developed in [BDK92] and a constraint deduction approach. The sequent calculus was implemented as the basis for deductive planning and plan reuse in SICSTUS PROLOG. First practical experiences demonstrated that the sequent calculus approach provides an efficient reasoning method when proofs are guided by tactics. The tactics support the declarative representation of control knowledge which helps to keep the search space to a manageable size. A tactic language is used to describe the tactics and makes it easy to develop and incorporate new tactics into the system.

A translation into constraint predicate logic is presented as an alternative approach. In this case, the undecidability of LLP reflects in the undecidability of constraint satisfiability, although the basic machinery of constrained resolution itself is known to be complete. This localization of the undecidability in the constraint part raises the hope of finding decidable fragments of LLP by isolating decidable fragments of the constraint theory.

Acknowledgements

We would like to thank Susanne Biundo, Hans-Jürgen Bürckert, Dietmar Dengler, Hans-Jürgen Ohlbach and Gert Smolka for their interest and for fruitful discussions.

References

[Aba89] Martin Abadi. The power of temporal proofs. *Theoretical Computer Science*, 65:35–83, 1989.

[Bau92] Mathias Bauer. An interval-based temporal logic in a multivalued setting. In D. Kapur, editor, *Proceedings of the 11th International Conference on Automated Deduction (CADE'11)*, LNCS 607, pages 355–369, Saratoga Springs,NY,USA, 1992. Springer.

[BBD⁺93] Mathias Bauer, Susanne Biundo, Dietmar Dengler, Jana Koehler, and Gabriele Paul. PHI - a logic-based tool for intelligent help systems. In *Proceedings of the 13th International Joint Conference on Artificial Intelligence*, Chambery, France, 1993.

[BDK92] Susanne Biundo, Dietmar Dengler, and Jana Koehler. Deductive planning and plan reuse in a command language environment. In *Proceedings of the 10th European Conference on Artificial Intelligence*, pages 628–632, 1992.

[Bib86] Wolfgang Bibel. A deductive solution for plan generation. *New Generation Computing*, 4:115–132, 1986.

[Bür91] Hans-Jürgen Bürckert. *A Resolution Principle for a Logic with Restricted Quantifiers*. Lecture Notes in Artificial Intelligence, vol. 568. Springer, 1991.

[Den94] Dietmar Dengler. An adaptive deductive planning system. In A. Cohn, editor, *Proceedings of the 11th European Conference on Artificial Intelligence*, pages 610–614, Amsterdam, NL, August 1994. John Wiley & Sons.

[FO92] Michael Fisher and Richard Owens. From the past to the future: Executing temporal logic programs. In A. Voronkov, editor, *Proceedings of the International Conference on Logic Programming and Automated Reasoning (LPAR'92)*, pages 369–380. Springer, Berlin, Heidelberg, 1992.

[Fri91] Alan M. Frisch. The substitutional framework for sorted deduction: Fundamental results on hybrid reasoning. *Artificial Intelligence*, 49:161–198, 1991.

[FS91] Alan M. Frisch and Richard R. Scherl. A general framework for modal deduction. In J. Allen, R. Fikes, and E. Sandewall, editors, *Proceedings of the 2nd Conference on Principles of Knowledge Representation and Reasoning (KR'91)*, pages 196–207, Cambridge,Massachusetts, 1991. Morgan Kaufmann.

[Gab89] Dov Gabbay. Declarative past and imperative future: Executable temporal logic for imperative systems. In H. Barringer and A. Pnueli, editors, *Proceedings of the Colloquium on Temporal Logic in Specification 1987*, LNCS 398, pages 402–450, Altrincham, 1989. Springer.

[Hal87] Roger Hale. Temporal logic programming. In A. Galton, editor, *Temporal Logics and Their Applications*, pages 91–119. Academic Press, 1987.

[Koe94a] Jana Koehler. An application of terminological logics to case-based reasoning. In J. Doyle, E. Sandewall, and P. Torasso, editors, *Proceedings of the 4th International*

Conference on Principles of Knowledge Representation and Reasoning, pages 351–362. Morgan Kaufmann, San Francisco, CA, 1994.

[Koe94b] Jana Koehler. Avoiding pitfalls in case-based planning. In *Proceedings of the 2nd International Conference on Artificial Intelligence Planning Systems*, pages 104–109, Chicago, IL, 1994. AAAI Press, Menlo Park.

[Krö87] Fred Kröger. *Temporal Logic of Programs*. Springer, Heidelberg, 1987.

[MM83] Ben Moszkowski and Zohar Manna. Reasoning in interval temporal logic. In E. Clarke and D. Kozen, editors, *Proceedings of the Conference on Logics of Programs*, LNCS 164. Springer, 1983.

[MW87] Zohar Manna and Richard Waldinger. How to clear a block: Plan formation in situational logic. *Journal of Automated Reasoning*, 3:343–377, 1987.

[Ohl91] Hans-Jürgen Ohlbach. Semantics-based translation methods for modal logics. *Journal of Logic and Computation*, 1(5):691–775, 1991.

[Rab69] Michael O. Rabin. Decidability of second-order theories and automata on infinite trees. *Transactions of the American Mathematical Society*, 141:1–35, 1969.

[RP86] Roni Rosner and Amir Pnueli. A choppy logic. In *Symposium on Logic in Computer Science*, Cambridge, Massachusetts, 1986.

[SH88] Andrzej Szalas and Leszek Holenderski. Incompleteness of first-order temporal logic with until. *Theoretical Computer Science*, 57:317–325, 1988.

[Sza86] Andrzej Szalas. Concerning the semantic consequence relation in first-order temporal logic. *Theoretical Computer Science*, 47:329–334, 1986.

[Sza87] Andrzej Szalas. A complete axiomatic characterization of first-order temporal logic of linear time. *Theoretical Computer Science*, 54:199–214, 1987.

[Tre92] Ralf Treinen. A new method for undecidability proofs of first order theories. *Journal of Symbolic Computation*, 14(5):437–457, November 1992.

[Wal89] Lincoln A. Wallen. *Automated Deduction in Nonclassical Logics*. MIT Press, Cambridge, London, 1989.

Towards First-Order
Concurrent METATEM

Mark Reynolds

Abstract. In [27] we applied the METATEM executable temporal logic programming language in a distributed system guise to a case study of a hospital patient monitoring system. The purpose was to test METATEM as a framework in which to implement prototypes for developing specifications of complex reactive systems. We demonstrated that it is indeed a very useful tool in such situations and that the framework has promising potential for further development.

In this paper, we summarize the lessons which were learnt for future developments paying particular attention to those aspects concerned with using full first-order temporal logic as the underlying language.

1   Introduction

This paper[1] is about ensuring adequacy of design of complex systems which carry on in on-going interaction with their environment. We will describe a framework in which a *prototype* can be built. This means a simplified, probably inefficient version of the system which can nevertheless (hopefully) be seen working. A prototype has two very important roles to play in the software and hardware development programme. For one it can be used to demonstrate to the client roughly how things will work. This may help justify that the formal specification does indeed capture the client's expectations. The other purpose is for the top-level implementer to test and try different methods of breaking up the task into smaller modular ones and become clearer in deciding what is required from each component.

We will use temporal logic as the formal specification language. See [22] for details.

---

[1] The author wishes to thank Marcelo Finger, Michael Fisher, Ian Hodkinson, Tony Hunter, Richard Owens and Ben Strulo for many helpful discussions. The work was supported by the U.K. Science and Engineering Research Council under the METATEM project (GR/F/28526).

Recently, temporal logic is also being used as a formalism in which to actually write programs. We see many examples of *executable temporal logic* from initial developments in [23] to the very complex METATEM framework we describe below and the temporal logic programming in [1]. The idea is simply that the program read as a sentence of a temporal language describes how it behaves when it runs. Thus executable temporal logics are *declarative* programming languages.

METATEM began with the simple "declarative past implies imperative future" idea in [15]. A complex expressive programming language was taking shape in [3] and an implementation of it was described in [9]. Recently we have seen it applied to modelling rail networks in [7] and developing into the concurrent programming language Concurrent METATEM (also called 'CMP') in [11] or [12]. As a higher level language it seems very appropriate for prototyping while lower level more efficient –and less expressive – temporal languages like TEMPURA ([23]) might be appropriate for final implementations.

In [27] we demonstrated the success of METATEM as such a prototyping language and saw just how well the various steps in the design programme mesh together when we use such a framework. There are great advantages of using the unified approach of the same temporal formalism for specification, implementation and formal justification.

However, METATEM, especially the first-order Concurrent METATEM, is still in its early stages of development so we expected many shortcomings. We learned about what aspects particularly need working on, and whether there are any more serious problems. In this paper, we summarize the lessons which were learnt for future developments paying particular attention to those aspects concerned with using full first-order temporal logic as the underlying language.

In [27], we chose a system development problem from the software specification and design literature and tried using the METATEM framework to take it through the stages of specification, implementation and justification. The example we chose was a patient monitoring system for an intensive care ward of a hospital from [28]. We describe the problem in section 3.

In section 2 we describe our temporal logic formalism. In section 4 we introduce the METATEM programming language. In section 5 we describe a realistic prototype implementation which we would want to be able to build when METATEM is better developed.

All the while we have been noting lessons for METATEM which we present in section 6.

2   Temporal Logic

We are going to be concerned with the behaviour of processes over time. Two very useful formal languages for describing such behaviour are the propositional temporal logic **PML** and the first-order temporal logic **FML** based on the temporal connectives until U and since S introduced by Kamp in [18]. The simpler propositional language allows us to express less and so is easier to deal with.

A crucial point in the executable temporal logic paradigm is that the same languages are used to specify the desired behaviour of a program and to actually write the program to satisfy the specification. In fact, in the ideal case, the specification and the program are the same thing.

In any case, amongst many other advantages, using the same language for specification and implementation gives us a head start in proving correctness of programs. After formally describing the kind of behaviours we are interested in and the languages for describing them, below we consider the appropriate temporal logics for doing such proofs.

2.1 Temporal Structures

The languages **FML** and **PML** are used to describe the behaviour of processes over time. The underlying flow of time will often be taken here to be the natural numbers – or equivalently some sequence $s_0, s_1, s_2, ...$ of states – but, in general, such temporal languages can describe changes over any linear order $(T, <)$ of time points.

In the propositional case the *state* at each time is just given by the truth values of a set $\mathcal{L}_P$ of atomic propositions or atoms. The behaviour we are describing is just the way the various atoms become true or false over time. To formalise this we use a map $\pi_P : T \times \mathcal{L}_P \rightarrow \{\top, \bot\}$ where $\pi_P(t, q) = \top$ iff the atom q is true at time t.

In first-order temporal structures the state at each time is a whole first-order structure with a domain of objects on which are interpreted constant symbols, function symbols and predicate symbols. Without any restrictions such situations would be too messy to describe formally so we make some assumptions. As described in [26] there are many sets of simplifying assumptions which can be made but the ones we make here are comfortable to work with and, at the same time, so general that other approaches can be easily coded in.

For a start we assume that each state is a first order structure in the same language. Further, although this is a real restriction, assume that we have no function symbols. This is not necessary for METATEM but we do not need function symbols in this case study and they just complicate the exposition. So suppose that $\mathcal{L}_P$ is a set of predicate symbols and $\mathcal{L}_C$ is a set of constants. We divide up $\mathcal{L}_P$ into a set $\mathcal{L}_P^n$ for each $n \geq 0$ being the n-ary predicate symbols.

We assume a constant domain $\mathcal{D}$ of objects but, over time, the extensions of the predicates change. To formalise this we use a map $\pi_P = \pi_P^0 : T \times \mathcal{L}_P^0 \rightarrow \{\top, \bot\}$ and a map $\pi_P^n : T \times \mathcal{L}_P^n \rightarrow \mathcal{D}^n$ for each $n = 1, 2, ...$ The interpretations of the constants is of course constant: we use a map $\pi_C : \mathcal{L}_C \rightarrow \mathcal{D}$.

In many of the definitions below we can include the propositional case as a special case of the first-order one by equating $\mathcal{L}_P$ with $\mathcal{L}_P^0$ and π_P with π_P^0.

2.2 Syntax

We use

- a countable set $\mathcal{L}_P^n$ of predicate symbols of arity n,
- a countable set $\mathcal{L}_V$ of variable symbols $x_1, x_2, ...,$
- and a countable set $\mathcal{L}_C$ of constant symbols $c_1, c_2, ...$

The terms of **FML** are just its variable and constant symbols. Recall that we are not considering function symbols in this paper.

The set of formulas of **FML** is defined by:

- if $t_1, ..., t_n$ are terms and p is an n-ary predicate symbol then $p(t_1, ..., t_n)$ is a formula,
- if α and β are formulas then so are $\top, \neg\alpha, \alpha \wedge \beta, \forall x\alpha, \alpha\, \mathcal{U}\, \beta$ and $\alpha\, \mathcal{S}\, \beta$.

We have the usual idea of free and bound variable symbols in a formula and so the usual idea of a sentence – i.e. a formula with no free variables. The class of formulas which do not have any variable symbols or constants form the well-formed formulas of the propositional language **PML**. In **PML** we only use 0-ary predicate symbols which are just propositions.

A formula of the form $p(\bar{u})$ is called a *positive literal*. A formula of the form $\neg p(\bar{u})$ is called a *negative literal*. A *literal* is either a positive one or a negative one. A literal is *ground* if it is also a sentence.

2.3 Semantics

A variable assignment is a mapping from $\mathcal{L}_V$ into $\mathcal{D}$. Given such a variable assignment V we assume it extends to all terms by just putting $V(c) = \pi_C(c)$ for any constant c.

For a temporal structure $\mathcal{M} = (T, <, \mathcal{D}, \{\pi_P^n\}_{n \geq 0}, \pi_C)$ a time point $t \in T$, a formula ϕ, and a variable assignment V, we define whether (or not resp.) $\phi(\bar{d})$ under V is true at t in $\mathcal{M}$, written $\mathcal{M}, t, V \models \phi$ (or $\mathcal{M}, t, V \not\models \phi$ resp.) by induction on the quantifier depth of ϕ.

Given some $\mathcal{M}$ and some ϕ, suppose that for all points t, for all formulas ψ of lesser quantifier depth than ϕ and for all variable assignments V we have defined whether or not $\mathcal{M}, t, V \models \psi$. Now for $t \in T$ and variable assignment V we define:

- $\mathcal{M}, t, V \models \top$.
- $\mathcal{M}, t, V \models q$ for a proposition q iff $\pi_P^0(t, q) = \top$.
- $\mathcal{M}, t, V \models p(\bar{u})$ for an n-ary predicate p and n-tuple $\bar{u}$ of terms iff $(V(u_1), ..., V(u_n)) \in \pi_P^n(t, p)$.
- $\mathcal{M}, t, V \models \neg\chi$ iff $\mathcal{M}, t, V \not\models \chi$.
- $\mathcal{M}, t, V \models \chi \wedge \psi$ iff $\mathcal{M}, t, V \models \chi$ and $\mathcal{M}, t, V \models \psi$.
- $\mathcal{M}, t, V \models \chi\, \mathcal{U}\, \psi$ iff there is $s > t$ in T such that $\mathcal{M}, s, V \models \psi$ and for all $r \in T$ such that $t < r < s$, $\mathcal{M}, r, V \models \chi$.
- $\mathcal{M}, t, V \models \chi\, \mathcal{S}\, \psi$ iff there is $s < t$ in T such that $\mathcal{M}, s, V \models \psi$ and for all $r \in T$ such that $s < r < t$, $\mathcal{M}, r, V \models \chi$.
- $\mathcal{M}, t, V \models \forall x\chi$ for $x \in \mathcal{L}_V$ iff for all $d \in \mathcal{D}$ $\mathcal{M}, t, W \models \chi$ where W is the variable assignment given by

$$W(y) = \begin{cases} V(y) & y \neq x \\ d & y = x. \end{cases}$$

It is easy to prove that the truth of a formula at a point in a structure does not depend on assignments to variables which do not appear free in it. So we can write $\mathcal{M}, t, v \models \phi$ where v is a partial assignment provided its domain does include the free variables of ϕ. When σ is a sentence – or a **PML** formula – we also write $\mathcal{M}, t \models \sigma$ iff $\mathcal{M}, t, \emptyset \models \sigma$ where $\emptyset$ is the empty map.

2.4   Models

Say that temporal structure $\mathcal{M}$ is a *model* of a sentence σ iff $\mathcal{M}, 0 \models \sigma$. A sentence is *satisfiable* iff it has such a model. A sentence σ is *valid* iff $\mathcal{M}, t \models \sigma$ for all structures $\mathcal{M}$ and for all time points t in $\mathcal{M}$.

Note that this means that we are using an *anchored* logic in the sense of [22].

2.5   Abbreviations

We read $\xi \mathcal{U} \psi$ as "ξ until ψ" and similarly for since. Note that our $\mathcal{U}$ is *strict* in the sense that $p \mathcal{U} q$ being true says nothing about what is true now. In some presentations of temporal logic, until is defined to be non-strict. We introduce an abbreviation $\mathcal{U}^+$ for non-strict until: $\xi \mathcal{U}^+ \psi$ iff $\psi \vee (\xi \wedge (\xi \mathcal{U} \psi))$. Confusingly, the METATEM programming language uses non-strict until and strict since as its basic operators.

As well as the classical abbreviations $\bot, \vee, \rightarrow, \leftrightarrow$ and $\exists$ we also have the following temporal ones:

$\bigcirc \phi$	" ϕ is true in the next state "	$\bot \mathcal{U} \phi$
$\bullet \phi$	"there was a last state and ϕ was true in the last state"	$\bot \mathcal{S} \phi$
$\bullet \phi$	"if there was a last state then ϕ was true in that state"	$\neg \bullet \neg \phi$
$\Diamond \phi$	" ϕ will be true in some future state"	$\top \mathcal{U} \phi$
$\blacklozenge \phi$	" ϕ was true in some past state "	$\top \mathcal{S} \phi$
$\Box \phi$	" ϕ will be true in all future states"	$\neg \Diamond \neg \phi$
$\blacksquare \phi$	" ϕ was true in all past states"	$\neg \blacklozenge \neg \phi$

There are also two more complicated abbreviations:

- $\xi \mathcal{W} \psi$ for $(\xi \mathcal{U} \psi) \vee (\Box \xi)$ meaning ξ stays true unless ψ becomes true and
- $\xi \mathcal{Z} \psi$ for $(\xi \mathcal{S} \psi) \vee (\blacksquare \xi)$ meaning ξ has been true since ψ was or since the beginning of time.

2.6   Separation

As one would expect from the declarative past /imperative future motivation, one distinction which plays an important role in METATEM is that between formulas which refer to the past and those which refer to the future. Let us make this precise.

A formula ϕ is a *not necessarily strict future time formula* iff it is built without $\mathcal{S}$. The class of *strict future time* formulas include only

- $\psi \mathcal{U} \chi$ where ψ and χ are both not necessarily strict future time formulas and
- $\neg \phi$, $\phi \wedge \psi$ and $\forall x \phi$ where ϕ and ψ are both strict future time formulas.

Dually we have strict and not necessarily strict past time formulas.

It is clear that a strict past time formula only depends on the past for its truth. This classification of formulas is the basis for Gabbay's separation property and separation theorem which is itself useful for establishing the expressive power of the METATEM language. See [14] for details which also include a discussion of the proof-theory of various temporal logics mentioned above.

3 Informal Description

In this paper, we will apply METATEM to a patient monitoring system (**PMS**) for use in the intensive care wards of hospitals. This problem was initially discussed in [28] but a fuller description appears in [6] which analyses it in terms of what they call the temporal-causal formalisation for specifications. In [27] we compare their approach with ours.

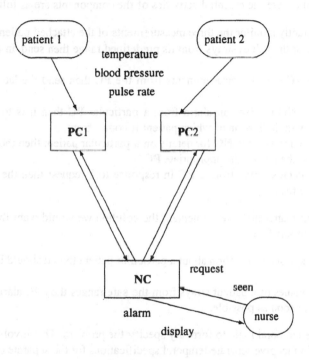

Fig. 8.1. A **PMS** with two patients

In the **PMS** there are two types of components: nurse components (**NC**'s) and patient components (**PC**'s). In any application of the system to a ward there will be a certain finite number of **PC**'s but only one **NC**. In the diagram we assume there are only two patients. Each **PC** is attached to a particular patient in the ward and automatically monitors various of her or his physiological measurements. For example, in this study we assume that blood pressure, pulse rate and temperature are measured. The **NC** is designed for use by the supervising nurse: the nurse can use the **NC** to find out information about a particular patient and the **NC** also displays an alarm if any of the measurements of any of the patients strays from a certain predefined range. Although this is not mentioned in the previous literature, we make things a little more interesting by assuming that only one set of measurements can be displayed on the NC at a time and that the nurse can indicate to the NC when she or he has seen the information and finished with it.

At the risk of already making decisions about the implementation of the problem rather than its specification we must say that the descriptions mentioned above do imply some form of distributed system with the separate components acting independently and communicating with each other rather than their being one centralised process. See figure 8.1. In that case the essential activities of the components are as follows:

PC: – constantly monitor the three measurements of the attached patient,
   – if any of the values strays from its predefined range then send an alarm to the NC,
   – if the NC requests measurements from the PC then send the latest measurements.

NC: – if the NC receives an alarm from a particular PC then it is to display the alarm with indication of which patient it comes from,
   – if the nurse asks the NC for details on a particular patient then the NC should request them from the appropriate PC
   – when details arrive from a PC in response to a request then the NC should display them.

In any case, no matter how we implement the solution we would want the following conditions to be satisfied:

   if the nurse requests information about a particular patient then it should be displayed and
   if any of the values of a patient strays from the safe ranges then the alarm should be activated as soon as possible.

In [27] we use temporal logic to formally specify the problem. This involves realising that we must in fact give separate temporal specifications for the separate components, describe the communication channels and specify which components are in charge of which channels. In [27], the above requirements are rendered as temporal formulae and we attempt to formally prove that the implementation satisfies them. In this paper we will not concentrate on the issues of the semantics of programming languages and modular proof theory which were raised by this attempt.

4 METATEM **Programming Language**

METATEM is really a paradigm for programming languages rather than one particular language. The bases are three:

   – programs should be expressed in a temporal language;
   – programs should be able to be read declaratively;

– the operation of the program should be interpretative with individual program
clauses operating according to the "declarative past implies imperative future" idea.

Most versions of METATEM use the temporal languages **PML** and **FML** with until and
since.

The basic idea of declarative languages is that a program should be able to be read
as a specification of a problem in some formal language and that running the program
should solve that problem. Thus we will see that a METATEM program can easily be
read as a temporal sentence and that running the program *should* produce a model of
that sentence.

The task of the METATEM program is to build a model satisfying the declared specification. This can sometimes be done by a machine following some arcane, highly
complex procedure which eventually emerges with the description of the model (see
[24]). That would *not* be the METATEM approach. Because we are describing a programming language, transparency of control is crucial. It should be easy to follow and
predict the program's behaviour and the contribution of the individual clauses must be
straight forward.

Fortunately, these various disparate aims can be very nicely satisfied by the intuitively appealing "declarative past implies imperative future" idea of [15]. The METATEM
program *rule* is of the form $P \Rightarrow F$ where P is a strict past-time formula and F is a not
necessarily strict future-time formula. The idea is that on the basis of the declarative
truth of the past time P the program should go on to "do" F.

In the case of a closed system, a METATEM program is a list $\{P_i \rightarrow F_i \mid i = 1, ..., n\}$
of such rules and, at least in the propositional case, it represents the **PML** formula

$$\Box \bigwedge_{i=1}^{n} (P_i \rightarrow F_i).$$

The program is read declaratively as a specification: the execution mechanism should
deliver a model of this formula. To do so it will indicate which propositions are true
and which are false at time 0, then at time 1, then at time 2, e.t.c. It does this by going
through the whole list of rules $P_i \rightarrow F_i$ at each successive stage and make sure that F_i
gets made true whenever P_i is. For details of the way it does this see [3].

4.1 Concurrency

The first complication we introduce to this situation is the very useful one of regarding
a METATEM process as a *Reactive Module*. This means that we immerse the process in
an environment. The basic idea of a reactive module, as described, for example, in [25]
or [16], is an object whose relationship with its environment changes over time in such
a way that part of the relationship is under control of the module and part under control
of the environment. The usual temporal logic approach to reactive modules is to posit a
strict division of the predicates or propositions of the language. There are "component
predicates" under the control of the module such as *alarm* e.t.c. mentioned above and
"environment predicates" such as *temp*. The special case in which all the predicates are
under the control of a module is often called a *closed system*.

Once we have the ability to define a reactive module it is not then a great step further to generalise to the idea of a distributed system with several modules acting concurrently. This distributed system may be closed or it may itself be immersed in an environment. The modules within the system are just the ordinary reactive modules we have described above but the environment of a particular module is now the extra-system environment in combination with all the other modules.

In order to implement a particular module we can use a METATEM program as described in the last subsection. Essentially, the program uses past truths about itself and its environment under the control of the "past implies future" rules as above to construct the present and future truths. In actual fact things are not quite so straight forward and there are several issues to clarify to do with the module communicating with its environment which we discuss below.

Another possibility for implementation follows when we see that a whole system of reactive modules is itself just a reactive module as well. In fact the possibility of a great hierarchy of modules opens up here. At its depths are atomic modules maybe implemented by METATEM processes but at all other heights a module at one level turns out to be a distributed system of modules when we look within it.

Now let us look at the arrangements for communication between modules. In CMP, we use broadcast message passing as outlined in [8]. This means that all announcements are potentially available for every module in the system to hear but each module has an abstract interface specification indicating which predicates it listens to in its environment and which it may announce. Broadcast message passing contrasts with various channel systems such as that in [17] in which messages pass from a specified sender to a specified receiver. The approaches are compared in [11].

We have the choice of a synchronous or asynchronous system. In the synchronous system, the overall controlling mechanism waits for all modules to be ready then lets them make their announcements simultaneously. Then each module is told about the announcements which its interface classes as inputs and its left to get on with its next step.

This is not how distributed systems usually operate though. Usually the modules are immersed in an asynchronous mileau. Here each module can make its announcements of the truth of its ouput predicates whenever it wants to. Each interested listener should be told then but this means that the listener might be interrupted in the middle of its computation of its next step. So instead the CMP system arranges for a message queue to be built up for each module containing, in order, the announcements of interest to that module, which have been made. Then, only when a module deigns to make an announcement – even an empty one – will it get access to its queue.

There are actually several options for the details of this access but an easily implementable arrangement may be for the process to read off the whole queue leaving it empty to collect messages announced during the next step. The module itself can then proceed, acting as if all the announcements contained within the queue happened yesterday. Notice that the ○ operator then loses its usual semantics. This also happens in different ways with the other options for reading through the queue.

One problem with this bunching arrangement is that if several identical messages arrive near to each other in the queue then they get collapsed into one message when we bunch. We thus lose count of messages while it is easy to think of situations in which this is important. Consider, for example, a module which is required to eventually announce a q for each time it hears a p. If it is delivered a message queue containing five p's then a rule $p \rightarrow \Diamond q$ will only produce one q. In Michael Fisher's CMP this problem is avoided by the module only reading off the queue up until the first repetition of messages. Of course, then the operational meaning of $\bigcirc$ reflects even less the semantics of yesterday. We will discuss this problem later when we will also note that a meta-language facility may also be of some help.

The existing version of propositional concurrent METATEM (Michael Fisher's 1992 CMP) approaches the question of control of propositions in a slightly different way than the strict division mentioned above. For a particular module there are *interface* propositions and these may either be in the *input* list, the *output* list or both. There are also *internal* propositions which are in neither list. There are really two slightly different temporal structures built by such a module.

The external structure is built in the language of the interface. Input propositions are made true when a broadcast message of that proposition is received by the module. Output propositions are made true when the module announces the truth of that proposition. Propositions in both lists are made true by either event. Thus we do not have strict separation of control of propositions.

The internal structure is built in the whole language of the module. Propositions are made true if either the module makes them true or they are in the input list and received by the module from an outside broadcast.

It is important to note that the external structure is not just a reduct of the internal structure to the language of the interface. There may be propositions in the head of clauses which are not in the output list of the module but are in the input list. Such a proposition may be true in the internal structure at exactly the same time as it is false in the external structure. In subsection 6.3 we discuss the suggestion in [11] that METATEM interpret such a rule, say $\bigcirc true \rightarrow p$, as "wait for p to be made true before proceeding". This can be viewed as a means of synchronising behaviour.

One consequence of the above procedure is that rules which purport to bring about the falsity of a proposition explicitly should not be taken too seriously if that proposition is in the listened-to set. The environment may make the proposition true anyway. For this reason we refrain from using negations in the future parts of rules in our case study. Given the "false unless otherwise stated" approach to recording truths, negations are redundant there anyway.

Note that in using METATEM to implement a reactive module, we can not rely on any back-tracking concerning announced truths. As described in [11], this means that some compromises have to made with the correctness of the module program as a declarative reading.

4.2 First-Order METATEM

Moving to consider programming in the predicate temporal logic **FML** we allow much more expressiveness and, unfortunately, find a whole host of new difficulties. Below we first introduce a predicate METATEM with very restricted syntax to show how the basic mechanism works and then consider relaxing some of the restrictions. This will explain the motivation for the long list of complicated conditions which we begin with.

Again we want the general "past implies future" format for rules but to keep tight control of the variables we require some more structure on each rule. In the *Separated Normal Form* (SNF$_f$) of [10] the program represents the formula

$$\Box \bigwedge_{i=1}^{n} \forall \overline{x_i}.(\forall \overline{y_i}.P_i(\overline{x_i}, \overline{y_i})) \rightarrow (\exists \overline{z_i}.F_i(\overline{x_i}, \overline{z_i}))$$

where each rule represents one of the following four types of formula:

An *initial* $\Box$ **rule:**

$$\forall \overline{x}.[(\forall \overline{y}. \bullet \bot \wedge \bigwedge_{b=1}^{k} l_b(\overline{x}, \overline{y})) \rightarrow \exists \overline{z}. \bigvee_{j=1}^{r} m_j(\overline{x}, \overline{z})]$$

An *initial* $\Diamond$ **rule:**

$$\forall \overline{x}.[(\forall \overline{y}. \bullet \bot \wedge \bigwedge_{b=1}^{k} l_b(\overline{x}, \overline{y})) \rightarrow \exists \overline{z}. \Diamond l(\overline{x}, \overline{z})]$$

A *global* $\Box$ **rule:**

$$\forall \overline{x}.[(\forall \overline{y}.(\bigcirc\!\!\!\!\bullet \bigwedge_{i=1}^{n} k_i(\overline{x}, \overline{y})) \wedge \bigwedge_{b=1}^{k} l_b(\overline{x}, \overline{y})) \rightarrow \exists \overline{z}. \bigvee_{j=1}^{r} m_j(\overline{x}, \overline{z})]$$

A *global* $\Diamond$ **rule:**

$$\forall \overline{x}.[(\forall \overline{y}.(\bigcirc\!\!\!\!\bullet \bigwedge_{i=1}^{n} k_i(\overline{x}, \overline{y})) \wedge \bigwedge_{b=1}^{k} l_b(\overline{x}, \overline{y})) \rightarrow \exists \overline{z}. \Diamond l(\overline{x}, \overline{z})]$$

where each k_i, l_i, m_j and l is a literal. It is shown in [10] that every sentence can be transformed into this form.

Unfortunately the full generality of this form is very difficult to deal with so we will also suppose that

- the left hand side of each rule is strict past – so $k = 0$,
- $\overline{y}$ and $\overline{z}$ are empty lists – i.e. all variables are one of the x_i,
- each variable symbol appears at least once in a positive literal on the left hand side of the rule.

Thus our syntax looks more like:

$$\bullet \bot \rightarrow \bigvee_{j=1}^{r} m_j(\overline{c})$$

$$\bullet \bot \rightarrow \Diamond l(\overline{c})$$

$$\forall \bar{x}. [\; \bullet \bigwedge_{i=1}^{n} k_i(\bar{x}) \rightarrow \bigvee_{j=1}^{r} m_j(\bar{x})]$$

$$\forall \bar{x}. [\; \bullet \bigwedge_{i=1}^{n} k_i(\bar{x}) \rightarrow \Diamond l(\bar{x})]$$

where each k_i, m_j and l is a literal and $\bar{c}$ is a tuple of constants.

Further assume that at any time the environment only announces a finite number of truths – as we've seen this is a sensible practical restriction. We will show that the process also only has to announce a finite number of truths at any stage. This is true at the beginning as the only firing rules are those of the form $\bullet \bot \rightarrow \bigvee m$ or $\bullet \bot \rightarrow \Diamond l$ where m and l are ground literals. Negative literals do not need to be announced.

Let us detail the basic recursive procedure now.

We suppose that as well as announcing truths at each stage the process stores the finite list of truths – component and environment truths – for that stage. Remember that positive ground literals are assumed false unless otherwise recorded.

As in the propositional case we have to, at each stage, determine which rules fire. Here it is slightly more complicated because we want to determine, for each rule,

$$\forall \bar{x}. (P(\bar{x}) \rightarrow F(\bar{x}))$$

which substitutions $[\bar{x} \mapsto \bar{d}]$ make $P(\bar{x})$ true now. It can be shown due to our assumptions about the form of rules and the finite extent of past truths that there are only a finite number of such substitutions. For each of these we just need to make $F(\bar{d})$ true and this done just as in the propositional case as the component has control of announcing truths involving any predicates in F.

Once again backtracking is an option but is not yet usually implemented. We saw that in the reactive module situation there are good reasons for not allowing backtracking anyway.

Why have we these restrictions and how could we lift them?

Having literals on the left hand side of rules that are not nested under past operators makes the computation procedure more difficult and yet given all the other restrictions we are imposing we do not seem to lose any expressive power outlawing them. In most cases it seems possible to rewrite the program equivalently without them.

Now consider the restriction that each variable symbol in a rule must appear in at least one positive literal on the left hand side. There are two sorts of examples where this does not hold. Consider first a rule like $\forall xy. [P(x) \rightarrow F(x, y)]$ in which a variable, here y, does not appear at all on the left hand side. This means that if $P(a)$ happens to hold sometime then the component will have to make $F(a, a)$, $F(a, b)$, $F(a, c)$, e.t.c. all true. This is clearly a serious problem for our general procedure if the domain is infinite: we may have an infinite set of eventualities to satisfy. Worse, for example if $F(x, y)$ is just a predicate, is the possibility that an infinite number of truths may need to be announced. Clearly this challenges our whole framework and has good reason not to be allowed.

Even if the domain is finite there is a not unsurmountable problem here in that the interpreter does not know what the domain looks like. In the case that the domain is fixed for all uses of the program this can be easily solved by converting the program into a propositional one instead. If the domain will always be finite but does vary from use to use of the program we might need some facility here for listing the domain.

Another example, which breaks the restriction is a rule like $\forall xy.\ \mathbf{O}\,(p(x) \wedge \neg q(y)) \rightarrow$ $R(x, y)$ in which y again does not appear in a positive literal on the left. Here, if $p(a)$ was true yesterday we need to find all the elements b of the domain for which $q(b)$ was not true yesterday. Again if the domain is finite and listed somewhere in the program this problem is solvable – we just look for those b for which $q(b)$ is *not* recorded holding yesterday – but if the domain is infinite we have the same seemingly insurmountable problem of sometimes having to announce an infinite number of truths.

Now let us consider allowing $\bar{y}$ to be nonempty i.e. allowing universally quantified variables within the left hand side of the rule. In these rules, such as $\forall x.(\forall y.\ \mathbf{O}\,q(x, y)) \rightarrow$ $F(x)$ we have the problem of not knowing the domain to contend with again.

Next let us consider allowing $\bar{z}$ to be nonempty i.e. allowing existentially quantified variables within the right hand side of the rule. For example, if $P(a)$ holds and we have a rule $\forall x.P(x) \rightarrow (\exists z.F(x, z))$ then the process has to find an element d such that $F(a, d)$ is true. If environmentally controlled predicates are involved in F in a certain way it may be that the environment, will, in its own good time supply the process with such a value – this could be a way of synchronously receiving a message. Similarly one could delay grounding the variable until a later stage when either the environment or component fixes its value. A different approach is to let the process to choose a new constant symbol or functional term instead. This solution may not be acceptable in the finite domain case.

To describe a module interface in the first-order case we just use a straight forward generalisation of that in the propositional case: predicates can be listened to and/or announced (or neither).

4.3   Program Syntax

In this subsection we introduce the syntax which we will use to implement our **PMS**. Consider, first the propositional case.

Strings consisting of lower case letters or a few symbols like - are used for proposition names. Call such a string an *atom*.

The logical operators are represented as follows:

$$\textbf{PML:}\quad \top \qquad \bot \quad \wedge \quad \vee \quad \neg$$
$$\textbf{CMP:}\quad \texttt{true}\ \texttt{false}\ \texttt{\&}\quad |\quad \sim$$

The temporal operators are:

$$\textbf{PML:}\ \mathcal{U}^{+}\ \mathcal{S}\ \bigcirc\ \mathbf{O}\ \bullet\ \Diamond\ \blacklozenge\ \Box\ \blacksquare\ \mathcal{W}\ \mathcal{Z}$$
$$\textbf{CMP:}\ \texttt{U}\ \ \texttt{S}\ \ \texttt{N}\ \ \texttt{Y}\ \ \texttt{Q}\ \ \texttt{F}\ \ \texttt{P}\ \ \texttt{G}\ \ \texttt{H}\ \ \texttt{W}\ \ \texttt{Z}\ .$$

It is straight forward to define the class *SPast* of strictly past expressions and the class *NFut* of not necessarily strict future expressions. A program rule looks like $\alpha => \beta$ for some $\alpha \in SPast$ and $\beta \in NFut$.

A module program body is just a list of rules $\alpha=>\beta$ separated by semi-colons and finished by a full stop.

The syntax of a module program must tell us about the control of propositions. The input or listened-to list is not present if its empty but otherwises is represented by $(p_1,...,p_n)$ where $p_1, ..., p_n$ are the input propositions. The output or announced list is also not present if its empty but otherwises is represented by $[p_1,...,p_n]$ where $p_1, ..., p_n$ are the output propositions.

In CMP we begin a module program with an atom being the module's name. The syntax is

$$< \text{module name} >< \text{input list} >< \text{output list} > : < \text{program body} >$$

To define a system of METATEM modules CMP uses the syntactic arrangement of listing their programs. The list ends in a full-stop after the last module program (which also ends in a full-stop). There is an option for selecting asynchronous or synchronous running of the modules.

When we present an **FML** program to implement the **PMS** in the next section we will just use a straight forward generalisation of the syntax for propositional CMP. The main differences will be that strings of lower case letters will stand for predicates (as well as propositions) and for constant symbols too (and of course module names). We will use the same upper case letters for temporal operators as before but as in PROLOG a string of letters beginning with an upper case one will be used to represent a variable. This overuse of case conventions should not be ambiguous there but any real first-order programming language will have to be more careful.

Variables will be assumed to be universally quantified across the whole rule (i.e. they are x_i types). In extensions of this language with less restrictions on the form of rules, some notation will have to be invented to distinguish other types of variables.

5 Program

In this section we dream. Forgetting the fact that METATEM is in its early stages of development we suppose that we have every reasonable facility that a METATEM prototyping system should have and present a possible program for implementing our **PMS** distributed system. Here we just note the new facilities that we are assuming of MET-ATEM and we discuss them in more detail in the next section. Recall that our task is to build a prototype which will establish a feasible specification for the separate modules.

Of course we can suppose that the running and debugging of the implementation is done in a nice package environment but we won't go into these aspects here as there are more fundamental decisions to be made about the operation of the framework first. However, we might consider interfering interactively in the running of the program by allowing the user to play at being the nurse. This seems possible to engineer but for now we will suppose that once again we want the system to to simulate the patients and nurse as well as running system components. Assuming two patients – Doris and Bahir– we have six components in this closed system.

One observation made on the propositional implementation made in [27] was that the bunching algorithm for the reading of message queues was not satisfactory: slower modules tended to get left far behind gathering very long message queues. In the program below we will assume that this algorithm is changed in exactly the way we want it to be: namely that the whole of a component's queue is emptied at each broadcast by the component.

We will also assume that the package allows the user to make her or his own choice about the relative firing rates of the components. Here, for example, we might want the patients' measurements to be made very often compared to the actions of the nurse. If such control is not given to the user then we could build in our own desired time-control mechanism into the program in several ways. One could be to introduce a new process called a clock which sends out "ticks". Then each process could be limited to only broadcast at certain multiples of "ticks". But this is all a lot of extra work and it seems a reasonable request to ask for user-controlled broadcast rates. Of course, this whole discussion shows that we again desire asynchronous behaviour by the different modules.

Enough then of the setting for the program. Now let us consider each process in turn. Recall that first-order METATEM has, at the moment, severe restrictions on the form of rules.

5.1 Program

The predicates used are 2-ary *temp*, 2-ary *blood*, 2-ary *pulse* under shared control of the various patients, 0-ary *seen*, 0-ary *act* and 1-ary *req* under the control of the nurse, 1-ary *alarm*, 4-ary *display* and 1-ary *pat-req* under the control of the NC, and 1-ary *pat-al* and 4-ary *pat-val* under shared control of the various PCs.

Doris: To allow the patient to broadcast a realistic range of measurements it seems best to suppose that some random-number generator is available. Without going into to the exact form of a simulator for fluctuations in the temperature (e.t.c.) of a critically ill patient – but, nevertheless noting that some research might need to be done by the implementor and some heavy duty mathematical functions might need to be provided in the METATEM language – we here assume that a program construct $RND(\delta)$ exists as a term where δ is the description of a probabilistic distribution. We suppose each time a formula containing the term is made true that the term gets substituted by a real or integer value according to the distribution δ. Recall that the three measurements are announced simultaneously.

```
doris[temp, blood, pulse]:
Y true => temp( RND(temp-distribution) );
Y true => blood( RND(blood-distribution) );
Y true => pulse( RND(pulse-distribution) ).
```

Bahir: As for Doris.

Nurse: We assume that the nurse acts on alarms and sees displays immediately. She or he will cycle through patients in some order requesting their information. It is likely that she or he wont make a request at each step but using a random-integer generator $RNDI(0, 9)$ with a similar behaviour to RND above, we can make her or him do so about every ten steps.

To produce the cycling in the order Doris, Bahir, Doris, Bahir, ... we assume that 1-ary *begin* and 2-ary *next* are some internal predicates. *begin(P)* will only ever be true at the first moment of time and then only with with *P = doris*. After that the nurse waits to request (*wtr*) information on doris until she or he does so when the right random number (here 0) turns up. Because of the constant extension of *next* she or he keeps then alternating between requesting each of the two patients.

Note that the nurse will never request information about a non-existent patient.

```
nurse(alarm, display)[act, seen, req]:
Y alarm(P) => act;
Y display(P,V) => seen;

Y begin(P) => wtr(P);
Y ( wtr(P) & ~req(P) ) => wtr(P);
Y( wtr(P) & req(P) & next(P,Q) ) => wtr(Q);

Y ( wtr(P) & ( RNDI( 0,9) =0 ) )
        => req(P);

Q false => begin(doris);
Y true => next(doris, bahir);
Y true => next(bahir, doris).
```

PC-Doris: To detect a life-threatening situation and send off a *pat-al* is straight forward. Note that here we need < comparison on values.

Answering requests is more of a problem for two reasons. First, our specification requires that when a request is received we send off values which are measured subsequently. To ensure that we do not send off old values we have to delay answering the request for at least a step and then send off the next values to arrive to make sure that they are "fresh". Thus the use of a proposition *reqd* which is made true by a request and stays true until values arrive and are sent off.

As we will see when we consider the **NC** component, it is crucial for our implementation that **PC**s only send off values when they are asked and only send off one set of values. This brings us to our next problem: it is possible that two (or more) sets of values arrive while the **PC** is making one step. Thus, if it has a request to answer, the **PC** has to choose one set to send. It seems that the restricted METATEM language for the first-order case is not adequate to implement this requirement.

As a possible solution to this problem we introduce a new 3-place auxiliary function *NDC* which enables such *non-deterministic choice*. The idea is as follows. Suppose that $\phi(\bar{y}, x)$ is a formula with free variables only from the set $\{x, y_1, ..., y_n\}$. In our program ϕ is only ever a positive literal but perhaps this need not be always so. Suppose further that $\bar{a}$ is an n-tuple of ground terms. Then the construct $NDC(X, \phi(\bar{a}, X), Y)$ can appear as a literal in the program for any variable symbols X and Y. To define its semantics suppose that at some time t, B is the set of all the ground terms which make $\phi(\bar{a}, b)$ true now. Then if B is empty we define $NDC(X, \phi(\bar{a}, X), d)$ to be false for all d at time t. On the other hand if B is not empty then the computer must choose one value $b \in B$ and at time t, $NDC(X, \phi(\bar{a}, X), d)$ is true if and only if $d = b$.

So in our program, if $temp(doris, 37.9)$ and $temp(doris, 43)$ are both assumed true at the same moment by **PC**-doris – no doubt because its running much slower than the thermometer – then its up to the program to pick one of 37.9 or 43 , say 37.9, and make $NDC(T, temp(doris, T), 37.9)$ true and $NDC(T, temp(doris, T), 43)$ false. We discuss this construct and other approaches in subsection 6.2.

Notice that the **PC** assumes that the three measured values come in simultaneously. Finally, consider the question of keeping $reqd$ true until a set of values comes in. We want to say that this happens if yesterday $reqd$ was true and it was not true that there is a value which came in. We see that we can use NDC to express this negated existential quantifier.

```
pc-doris(temp, blood, pulse,patreq)[patal, patval]:


Y( temp(doris,T) & (T< mint) ) => patal(doris);
Y( temp(doris,T) & (T> maxt) ) => patal(doris);
Y( blood(doris,B) & (B< minb) ) => patal(doris);
Y( blood(doris,B) & (B> maxb) ) => patal(doris);
Y( pulse(doris,R) & (R< minr) ) => patal(doris);
Y( pulse(doris,R) & (R> maxr) ) => patal(doris);


Y( patreq( doris) ) => reqd;


Y( reqd & ~NDC(T, temp(doris,T), T) ) => reqd;

Y( reqd & NDC(T, temp(doris,T), T))
   & NDC(B, blood(doris,B), B))
   & NDC(R, pulse(doris,R), R))
      => patval(doris,T,B,R).
```

PC-Bahir: As for **PC**-Doris.

NC: Here we use the *begin, next* pattern as in the nurse's program to allow the **NC** to go through the list of patients satisfying requests for information. This is because we only want to send out one *pat-req* at a time and be waiting for one reply. Meanwhile the nurse might be making many varied requests. These are remembered in a simple yes/no fashion by setting $reqd(P)$ to be true when a request for P comes in and keeping it true until the **NC** sends out a *pat-req(P)*. Then it makes $wait(P)$ true and keeps that true until a set of values comes back. Then that is displayed until the nurse has *seen* it which signals the **NC** to start checking through other requests.

Notice that we are having to use the new predicate $reqd$ where, in the propositional program, we were able to use a since expression. In order to satisfy the specification we may only cancel a request when we send off a *pat-req*. If we ignored requests for P coming in during a wait for P we may end up displaying "old" information gathered before the request but displayed after it. To avoid this we have to process such a subsequent request at a later time.

Notice that instead of using a negated *NDC* to express $\neg\exists$ as in the program for **PC-doris** we bring about the end of a waiting period by using the *stop–wait* proposition. This is an acceptable method of expressing $\neg\exists$ when we can delay the effect for one step.

In section 11.3 we discuss a completely different way of arranging for the **NC** to eventually deal with requests.

```
nurcomp(patal,patval,seen,req)[patreq,alarm,display]:

Y patal(P) => alarm(P);

Y begin(P) => chance(P);

Y( req(P) ) => reqd(P);
Y( reqd(P) & ~chance(P) ) => reqd(P);

Y( chance(P) & reqd(P) ) => wait(P);
Y( chance(P) & reqd(P) ) => patreq(P);

Y( wait(P) & ~ stop-wait(P) ) => wait(P);
Y( wait(P) & patval(P, T,R,B) ) => display(P, T,R,B);
Y( display(P,T,R,B) & ~ seen ) => display(P, T,R,B);
Y( patval(P,T,R,B) ) => stop-wait(P);

Y( display(P,T,R,B) & seen & next(P,Q) ) => chance(Q);

Q false => begin(doris);
Y true => next(doris, bahir);
Y true => next(bahir, doris).
```

We must also remember to finish off the whole program with a full-stop.

6 Lessons for METATEM

During the course of the case study in [27] we saw a propositional version of the problem implemented quite acceptably by CMP.

However, in describing the problem in its full predicate guise we discovered several aspects of the existing METATEM language which did not seem correctly defined for our purposes and many many examples of features which would need to be added to form a realistic prototyping framework. Let us look in some detail at these points.

As listed in [27], we would want a few built-in arithmetic functions and relations, *typing* of objects in the object domain and so types of variable symbols (in the usual manner of abstract data types for object oriented languages [2], [20]), and a comfortable environment in which to program, observe and debug.

6.1 Rigid Predicates

We saw in our implementation that we sometimes want to be able to define certain predicates to have constant extension. In the NC for example, it would have been convenient to declare *next* to be such a *rigid* predicate. In the implementation we achieved the rigidity of *next* (except at the first step when it did not matter) by having its extension computed every step. This seems rather wasteful and we can easily imagine other, far worse, situations in which such a predicate has a very large extension or even an infinite one such as list membership has. An efficient METATEM implementation should allow us not to have to calculate such extensions each time. Fortunately, this will be easy to accommodate.

Although none appeared in our example, there is a related class of predicates which are also worth considering here: namely *derived* predicates. This is a classification which, as we show below, can be introduced to capture the idea that sometimes there is a "rigid" logical relationship between various flexible predicates which remains fixed as they evolve through time. For the sake of efficiency, or even just implementability, it may make sense to exploit this relationship.

We will introduce a modification to METATEM which will allow us to do this by classifying predicates into *basic* and *derived* and show that rigid predicates can be efficiently implemented within this framework as just a special type of derived predicate.

For now just suppose that our METATEM rules have single predicates as their heads. We syntactically divide the predicates up into derived ones and basic ones and we divide the module program up into a derived part which looks like a PROLOG program and a basic part which uses our restricted METATEM syntax. In the basic part we bar derived predicates from the heads of rules and in the derived part we bar basic predicates from the heads of clauses.

We only record the truth of basic positive ground literals in the historical database. To determine the truth of basic ground literals at a certain time we look up the database. To determine the truth of derived literals we use the usual PROLOG resolution procedure in the derived part of the program.

For example, consider the program line

```
Y( display(P,T,R,B) & seen & next(P,Q) )=> chance(Q);
```

when *next* is declared derived and the rest of the predicates basic. Finding *display*(*doris*, 38, 70, 120) and *seen* were true yesterday, say, we come to trying to find Q such that *next*(*doris*, Q) was true. Instead of looking in the database for yesterday's truths we go to the derived part of the program, namely

```
next(doris, bahir).
next(bahir, doris).
```

and using PROLOG's resolution strategy quickly find $Q = bahir$ is a solution. In keeping with our semantics we would have to try for other solutions too but finding none the NC can just announce *chance*(*bahir*).

As another example consider an implementation of the PMS which allows us to admit and discharge patients. Suppose that we have a basic predicate *patlist* which is true of the list of patients each time. Say *patlist([bahir, jim, antje])* was true yesterday. Now the derived program may, in part, look like:

```
next( X, Y) :- patlist( [H |T ]), nextl(H, [H|T], X, Y).
```

```
nextl( F, [X| [Y|T] ], X, Y).
nextl( F, [X], X, F).
```

When asked about *next(jim, Q)* being true yesterday, this should tell us that *Q = antje* is the only solution.

Notice that we could have used a more METATEM-like syntax here and written

```
patlist( [H|T] ) & nextl( H ,[H|T], X, Y)
  => next( X, Y);
```

e.t.c.

There are questions to be answered about allowing temporal operators into derived part of the program but even without allowing this we see that we have been able provide a not too messy means of increasing expressiveness and efficiency.

It may be thought that allowing other modules to have access to the "definitions" contained in the derived part of a module's program might save some repetition of complicated code. This is true but detracts from the realism of the simulation of a distributed system in which all information should be passed through the proper channels.

6.2 Basic Expressiveness

In the implementation we saw that we were able to express a remarkable variety of temporal relationships with just a very restricted syntax of program rule. Generally our rules were of the form $\bullet \perp \rightarrow p(\bar{c})$ or $\forall \bar{x}.[\ \bigcirc(C) \rightarrow p(\bar{u})]$ where p is a predicate, $\bar{c}$ a tuple of constants, C a conjunction of literals and $\bar{u}$ a tuple of terms. Thus the program was almost totally deterministic.

The only real deviation from this pattern was our need to introduce the non-deterministic choice operator *NDC*. Although this has a reasonable, if not completely simple semantics, it might be worth considering some other solutions to the problem. Recall that the problem is one of choosing just one from a whole host of values which might have made a predicate true yesterday. In the next subsection we see that waiting to synchronise with an incoming message presents another solution while in the following subsection we recall that changing the message queue bunching algorithm provides a yet different solution. Now let us look at some variations on the theme of non-deterministic choice.

One way which provides a more low level control of the operation of *NDC* is to introduce set operations into the basic repertoire of the program and allow non-deterministic choice of elements from a set. For example, allow a basic relation *in* to be the element-of relation between objects and sets and introduce *NDCS(element, set)* as the operator. Our **PC**-doris program can now contain the line

```
Y( temp(doris, T) ) => T in Tempset
```

e.t.c. to collect values in sets and the line

```
Y( reqd & NDCS(T,Tempset)
         & NDCS(B, Bloodset)
      & NDCS(R, Pulseset) )
 => patval( doris,T,B,R)
```

to select values and send them off. This construction seems eminently implementable if we promise to stick to finite sets.

As pointed out by Michael Fisher, in [13], using NDCS(e, S) is much like using an existential quantifier $\exists e \in S$. However, we have a very restricted version in terms of scope and temporal extent i.e. you just pick an element out of a pre-existing finite set and there is a guarantee that we do not have to go on computing with it and later try to back-track.

We saw in the implementation for **PC**-doris that the *NDC* operator was useful in expressing a negated existential quantifier within the body of a rule. Although our implementation for the **NC** showed that this is not necessary, we note that the alternative method used there – via the *stop – wait* proposition – required an extra step to stop the waiting. It is conceivable that an extra step may sometimes be crucial and so we can conclude that in the absence of any type of non-deterministic choice, some other immediate "not exists" operator might be called for.

There were other situations where non-determinism might have been useful. These include the scheduling of answers to requests by the **NC**. It would have been possible to to write something like

```
Y( req(P) ) => F patreq(P)
```

along with conditions to stop more than one request going out at once. Thus we could use METATEM's memory for obligations on the future to force that the *pat-req* is eventually made. This saves us cycling through the patients and may involve some use of a non-deterministic choice by the METATEM interpreter itself when it comes to try to satisfy several incompatible eventualities.

Another place where eventualities would have been useful is arranging for a **PC** to answer a request. We discuss this in the next section.

6.3 Synchronising Communications

Many distributed systems rely to some extent on synchronous message passing – even in an asynchronous mileu. We need to supply the CMP language with some such facility. There is some discussion of this in [11] where it is suggested that allowing environment controlled predicates in the heads of rules might be a way of achieving this. If *p* is controlled by the environment, and the module finds in the head of a firing rule that it must make $p(X)$ true, for example, then the module must wait until it hears $p(a)$ from the environment for some *a* and immediately the module goes ahead with its step and broadcasts.

A problem with this approach is that it is not clear what to do when the module is waiting for two different broadcasts at once but receives them at different times.

However, as suggested by Michael Fisher in [13], this problem can be avoided if we restrict the syntax so that only one synchronization message is awaited at any one time, or, at worst, the module must wait for a disjunction of possible messages.

Still, it seems to be worth further investigating this approach. Alternatives might include built-in meta instructions to wait and synchronise with a specific other module. A related communications question concerns the use of channels for message passing between specific modules as opposed to broadcast message passing. This is discussed in [11] but it seems that there is much work still to be done on providing CMP with a realistic range of these facilities.

One use we could find for the method of synchronisation described above is a little trick so that we do not need non-deterministic choice to implement **PC**-doris. If we use

```
Y( patreq(doris) ) => temp(doris, T)
& blood(doris,B) & pulse(doris,R) & patval(doris, T,B,R).
```

to effect answering of requests then, after a request, **PC**-doris will wait for the very next broadcast of values before immediately sending off a *patval*. As doris only broadcasts one set of values at a time, this will allow us to satisfy the problematical requirement that **PC**-doris only send one answer to a request. Given the problems with this approach and the fact that the synchronisation is only used here incidentally, we decided not to use this code.

A superficially similar approach to dealing the request answering problem is that implied in the following code

```
Y( patreq(doris) ) =>
   F(   temp(doris, T)
      & blood(doris,B)
    & pulse(doris,R)
 & F( patval(doris,T,B,R)).
```

This is self explanatory because it is very close to our specification. The code also has a straight forward procedural interpretation which, it will be noted, involves no use of synchronicity. However, it is not possible to write such code in the restricted syntax which were using. Furthermore, if such code was possible to use we would have to be very careful about the bunching of messages to avoid running into the same old non-determinancy problems. This brings us to look at bunching.

6.4 The Message Queue

From the simple propositional implementation of the **PMS** in [27], one of the main lessons to be learnt was that the existing algorithm for bunching of messages in the message queue was unsuitable for our simulation purposes. Slower modules would tend to get left further and further behind dealing with messages from the distant past and leaving a longer and longer queue still to be dealt with. Here we will examine the alternative algorithms for message reading and see if we can suggest a version which will be generally applicable.

There is a whole range of possible message bunching algorithms. At one extreme is the "whole queue" algorithm which we assumed for developing the predicate version of an implementation: when a module broadcasts it is given the whole of its message queue to process in the next step. At the other extreme is a "one at a time" bunching algorithm in which the module is only given the first message from the queue (i.e. the oldest) to deal with. In between are many other algorithms including the one used in the existing CMP implementation which reads up until the first repeated message or takes the whole queue if there are no repeated messages.

We argue that any algorithm apart from the "whole queue" one suffers from the serious problem described above. Having slow modules being left behind is unrealistic and also seems to present a problem for the formal temporal semantics of CMP. One obvious problem is that the yesterday or last-time operator completely loses its usual semantics. Instead we suggest that the "whole queue" algorithm is used and if it is desired that a module should work slowly through its message queue then the module program should say this explicitly.

Now let us consider the disadvantages of this choice and if there are ways around them. There seems to be two main reasons for wanting to use a different algorithm. They are

- the inability to *count* the number of separate broadcasts of identical messages which arrive in one bunch and
- the inability to discern the temporal *ordering* of any of the messages which arrive in one bunch.

It seems that both of these problems come closer to solution if at each broadcast we deliver to the module the whole queue in the form of a list of messages in order rather than as a set. The problem then becomes one of providing the module program with enough machinery to adequately deal with such a list.

One unsatisfactory way of doing this is to use meta-language facilities to, if that is so desired for a particular module, lead the module through the queue as if the messages were arriving one at a time or in small packets defined in some other way. Thus can we simulate other bunching algorithms by using meta-language. This is unsatisfactory because the meta-language is (arguably) in general much less efficient as a programming language and the formal semantics are also much more complex.

It is much better to provide the necessary machinery at the object level. This needs much more work but as the message list is just a piece of temporal information – the list in order of broadcasts between the last two module ticks – it must not be too difficult to express properties of the list within the bodies of rules. In fact, by using an appropriate predicate and assuming the natural numbers are in the object domain, it is even possible to count occurrences of the same message.

One suggestion is to step back from the very restricted syntax written as $Y(B_1 \& ... \& B_n)$ involving a (supposed) yesterday operator and allow a more semantically justifiable use of S in combination with a special "tick" proposition.

6.5 Control of Predicates

In [27] we saw that the issue of control of predicates was very important in several aspects of our use of CMP. These included:

- the procedure for the operation of broadcast message passing,
- the question of implementability of specifications and
- modular proofs of correctness of programs with respect to specifications.

While it is essential for the operation of the module that some information is declared in the syntax of the module for the message passing procedure we argue that there are also good reasons to include some explicit declaration of control for the purpose of correctness.

We also saw that in some cases, for the purposes of correctness of the overall system, we might want to declare general information about control of a predicate. For example we might want to say that $pat - val$ is only controlled by modules of type **PC**. This type of information more properly belongs to the system as a whole rather than a particular module. It certainly becomes important if the system of modules is treated as a module itself

Some work needs to be done on the syntax for declaring all this sort of information. This includes the cases where joint control of predicates can be more specifically defined in a syntactical way. We saw an example of this where $pat\text{-}val(P_0, T, B, R)$ is only controlled by module **PC**-P_0.

7 Conclusion

We saw very good reason throughout the study reported here and in [27] to assume that given those particular improvements identified, the whole process of specification, implementation and justification would have proceeded remarkably well. We saw how METATEM would provide a very comfortable unified framework for each of the stages in the process.

Furthermore, we saw very good reasons why one would expect many of these improvements to be achieved without too much difficulty. None of them seemed to imply any great technical difficulty. Thus, in a few years, a similar case study should be able to present a very different, much more successful account of the development of the **PMS**.

Some of the problems identified seem very straight forward problems which must have been solved many times before in other contexts. These include supplying META-TEM with a comfortable programming and observing environment, adding PROLOG-like rigid predicates, and adding some kind of non-deterministic choice operator.

Some of the problems which seem likely to require a little more thought or policy decisions include allowing for synchronous communications, waiting for an environment input in other ways, handling the message queue and tidying up the interface definitions.

In the more distant future some thought is going to have to be applied to allowing for a hierarchy of modules and dynamic changes to the system. This study did not have much to say about these problems.

In short we have identified many directions in which work must be done but have seen that a very successful development is promised and does not seem too far away.

References

1. M. Abadi and Z. Manna. Temporal logic programming. *Journal of Symbolic Computation*, 8:277 – 295, 1989.
2. M. Atkinson, F. Bancilhon, D. DeWitt, K. Dittrich, D. Maier, and S. Zdnik. The object-oriented database system manifesto. In *[19]*, pages 223–240.
3. H. Barringer, M. Fisher, D. Gabbay, G. Gough, and R. Owens. METATEM: A framework for programming in temporal logic. In *Proc. of REX Workshop: Stepwise Refinement of Distributed Systems – Models, Formalisms and Correctness*, pages 94–129. LNCS Vol 430, Springer-Verlag, 1990.
4. H. Barringer. The use of temporal logic in the compositional specification of concurrent systems. In A. Galton, editor, *Temporal Logics and Their Applications*. Academic Press, 1987.
5. C. Beeri. Formal models for object-oriented databases. In *[19]*, pages 405–430.
6. J. Castro and J. Kramer. Temporal-causal system specifications. In *Proceedings of IEEE. International Conference on Computer Systems and Software Engineering (CompEuro90)*, May 1990.
7. M. Finger, M. Fisher, and R. Owens. METATEM at Work: Modelling Reactive Systems Using Executable Temporal Logic. In *Sixth International Conference on Industrial and Engineering Applications of Artificial Intelligence and Expert Systems (IEA/AIE-93)*, Edinburgh, U.K., June 1993. Gordon and Breach Publishers.
8. M. Fisher and H. Barringer. Concurrent METATEM Processes - A language for distributed AI. In *Proceedings of the European Simulation Multiconference*, June 1991.
9. M. Fisher and R. Owens. From the Past to the Future: Executing Temporal Logic Programs. In *Proceedings of Logic Programming and Automated Reasoning (LPAR)*, St. Petersberg, Russia, July 1992. (Published in *Lecture Notes in Computer Science*, volume 624, Springer Verlag).
10. M. Fisher. A Normal Form for First-Order Temporal Formulae. In *Proceedings of Eleventh International Conference on Automated Deduction (CADE)*, Saratoga Springs, New York, June 1992. (Published in *Lecture Notes in Computer Science*, volume 607, Springer Verlag).
11. M. Fisher. Concurrent METATEM — A Language for Modeling Reactive Systems. In *Parallel Architectures and Languages, Europe (PARLE)*, Munich, Germany, June 1993. Springer-Verlag.
12. M. Fisher. A Survey of Concurrent METATEM — The Language and its Applications In *Proceedings of First International Conference on Temporal Logic (ICTL)*, Bonn, Germany, July 1994. (Published in *Lecture Notes in Computer Science*, volume 827, Springer Verlag).
13. M. Fisher. Private Communication, October 1994.
14. D. Gabbay, I. Hodkinson, and M. Reynolds. *Temporal Logic: Mathematical Foundations and Computational Aspects, Vol. 1*. Oxford University Press, 1994.
15. D. M. Gabbay. Declarative past and imperative future: Executable temporal logic for interactive systems. In B. Banieqbal, H. Barringer, and A. Pnueli, editors, *Proceedings of Col-*

loquium on Temporal Logic in Specification, Altrincham, 1987, pages 67–89. Springer-Verlag, 1989. Springer Lecture Notes in Computer Science 398.

16. D. Harel and A. Pnueli. On the development of reactive systems. Technical report, Technical Report CS85-02, Department of Applied Mathematics, The Weizmann Institute of Science, Revohot, Israel, January 1985.

17. C. Hoare. *Communicating Sequential Processes*. Prentice-Hall, 1985.

18. J. Kamp. *Tense Logic and the theory of linear order*. PhD thesis, Michigan State University, 1968.

19. W. Kim, J. Nicolas, and S. Nishio, editors. *Proceedings of International Conference on Deductive and Object-Oriented Databases, Kyoto, Japan*. North-Holland, December 1989.

20. W. Kim and F. Lochosky, editors. *Object-Oriented Concepts, Databases and Applications*. ACM Press, Addison-Wesley, 1989.

21. L. Lamport. Proving the correctness of multiprocess programs. *IEEE. Trans. on Software Engineering*, SE-3(2):125–143, 1977.

22. Z. Manna and A. Pnueli. *The Temporal Logic of Reactive and Concurrent Systems: Specification*. Springer-Verlag, New York, 1992.

23. B. Moszkowski. *Executing Temporal Logic Programs*. Cambridge University Press, 1986.

24. A. Pnueli and R. Rosner. On the synthesis of a reactive module. In *Proceedings of the Sixteenth Symposium of Principles of Programming Languages*, pages 179 – 190. ACM, 1989.

25. A. Pnueli. *Applications of Temporal Logic to the Specification and Verification of Reactive Systems: A. survey of current trends*. Number 224 in Lecture Notes in Computer Science. Springer-Verlag, August 1986.

26. M. Reynolds. Axiomatising first-order temporal logic: Until and since over linear time. Technical report, Imperial College, 1992. Submitted to *Notre Dame J. Formal Logic*.

27. M. Reynolds. METATEM in intensive care. Technical report, Imperial College, 1993.

28. W. Stevens, G. Myers, and L. Constantine. Structured design. *IBM. systems Journal*, 13(12):115–139, 1974.

Solving Air-Traffic Problems with "Possible Worlds"

Marcos Cavalcanti

Abstract. We present here an executable modal logic based system: PW-$XRete$[1]. This system is connected with the modal logic through the Kripke's *possible worlds semantics*. PW-$XRete$ presents a procedure of *labeling the worlds* that provides an efficient implementation of the possible worlds and, as it showed, is well suited for nonmonotonic reasoning.

As an example of the use of PW-$XRete$ in real life situations we present its solution to the aircraft sequencing problem.

1 Introduction

The *real time expert systems* interact with outside world and work with incomplete datas. The incomplete characterization of the real world impose a *revocable* reasoning: the resulting deductions can be changed by the arrival of new informations.

In the last years many systems based on *hypothetical reasoning* were developed. Most of them employed:

- *logical level*: the stable models semantics [Fag91] and nonmonotonic logics (as default logic), to take into account the nonmonotonic character of the reasoning.
- *algorithmical level*: either the *TMS* [2], which utilise nonmonotonic inference mechanism and work in an unique context; or the *ATMS* [3], that examine in parallel all the possible completions of current world's description and that, in its original version, use a monotonic reasoning.

The system presented here, PW-$XRete$, is a *non-determinist programming language* addressed to realise real time expert systems. It allows a concurrent work in many worlds and uses a nonmonotonic inference mechanism.

---

[1] Possible Worlds XRete.

[2] Truth Maintenance Systems.

[3] Assumption-Based Truth Maintenance Systems.

This article is organised in the following way. In the next section we present $\mathcal{PW}$-*XRete*; in the section three, we show the connection between $\mathcal{PW}$-*XRete* and the modal logic; section four presents a real industrial application (the aircraft sequencing problem - ASP) and section five closes the paper with our conclusions and a comparison with related works.

2 $\mathcal{PW}$-*XRete*

As all production systems, $\mathcal{PW}$-*XRete* is composed by a finite set of objects, the *working memory* (WM), and a set (RB) of production rules like:

IF (conditions) THEN (conclusions).

The conditions are related to the working memory (WM) and the conclusions concern the deducted facts and the created worlds. A rule like

```
rule r1
    (a ?x ?y)
    (b ?x)
    →
    assert(c ?x ?y)
endrule;
```

can be seen as a logical implication where the variables are universally quantified:

$$\forall x, \forall y, a(x, y) \wedge b(x) \Rightarrow c(x, y).$$

The inference mechanism (IM) is composed by the four following steps:

Fig. 9.1. $\mathcal{PW}$-*XRete*'s Inference Mechanism

- Pattern-matching - Determines the set of *satisfied rules* (the rules' instances)[4].
- Conflict resolution - The choice of a rule among the set of satisfied rules[5].
- Firing rules - The firing of a rule execute the actions defined in the right hand side of the rule.
- Communication - It permits the interaction with the user.

2.1 Conservative extension of a production system

The advantages of the production systems in relation to the classical programming languages are well known: modularity, declarativity and independent rules[6].

PW-XRete has the particularity of been a production system that works in many *worlds* at a time.

Worlds The idea behind the semantics of the possible worlds (PW) presented by Kripke is to use a first order logic models as PW to the modal logic[7]. Thus, *each world in PW-XRete is a first order logic* **model**.

In the classical logic the truth value of a well formed formula (wff) never changes. With the notion of *possible worlds* (PW) the truth value of a wff *depends on the world in which it is placed*.

An *interpretation* is no more a unique set of objects, functions or relations as it is in the classical logic, but it is given by such a set *in each PW*. A wff ϕ is true in a w_i world if and only if (IFF) it is true according according to the classical interpretation associated to w_i. More formally:

Definition 2.1 *Let <PW, R> be a modal frame[8]. An* **interpretation** *in <PW, R> is a function π such as:*

$$\pi : PW \times VP \rightarrow \{true, false\}$$

where VP is the set of propositional variables of the language.

An interpretation indicates if a proposition P (in VP) is true in a w_i world (in PW). It should be remarked that such a definition of interpretation *extends* the classical one taking into account the world in which the formula is considered.

Contrarily to the classical logic, we can have a proposition with more than one truth value, one in each world.

Example: Let PW = {today, tomorrow} and VP = $\{P_1\}$

$\pi(\text{today}, P_1) = \text{true}$

---

[4] It's the most expensive operation of the IM. According to Rousset & al. ([RFC85]), the pattern-matching represents about 60% of the execution time.

[5] This choice can be made according to different strategies: we can have a completely non-determinist choice, a choice respecting a given priority, etc..

[6] For more details, see [DK77] and [Gab85].

[7] [Kri59a], [Kri59b].

[8] *PW* represents a non-empty set of possible worlds and *R* the accessibility relation among the worlds.

π(tomorrow, P_1) = false

Accessibility relation: In $\mathcal{PW}$-*XRete* the semantics of the accessibility relation is the same given by Kripke [Kri63]:

Definition 2.2 *Given two worlds* w_1, $w_2 \in PW$, *we say that* w_2 *is in relation with* w_1 *(written: (w_1 **R** w_2)) IFF all truth proposition in* w_2 *is at least possible in* w_1.

But different properties of the accessibility relation give different modal logics [Fro91]. We should thus define the *characteristics* of the worlds' accessibility relation in $\mathcal{PW}$-*XRete*.

Properties of the accessibility relation: In $\mathcal{PW}$-*XRete* the accessibility relation among the worlds is a *tree relation*. It is:

- *reflexive:* Each $w_i \in PW$ is accessible to itself, i.e., each truth proposition P in w_i is also possible in $w_i{}^9$;
- *transitive* IF (w_1 **R** w_2) and (w_2 **R** w_3) THEN (w_1 **R** w_3).

The last but not the least remark is that the accessibility relation defines a nonmonotonic logic IFF it is *well-founded*, i.e., if there is no infinite sequence l_1, l_2, ... such that ... $\preceq i_2 \preceq i_1$ [Bel90].

Our accessibility relation is defined as a tree relation where the *minimal element* is the *root r:* it is a *well-founded partial ordering relation*.

Labels and Objects The objects in $\mathcal{PW}$ – *XRete* are defined by a *class*, a list of *attributes* and a *label*[10].

Example$_2$: The condition

$$\{(\text{plane id=1 cat=2})\} \text{ in } ?w0$$

represents an object of the class *plane* with the attributes *id* et *cat* which identifies the plane and indicates its category, respectively. The world's variable *?w0* will be instantiated with one of the worlds in which this object is true.

$\mathcal{PW}$-*XRete* employs the idea of *viewpoints* present in Art [Art86]. The intention is to take into account the *nonmonotonic justifications in the multiple worlds*.

The label of a fact is thus defined by the *minimal world* (the most general) in which the fact has been asserted (we will call this world *"assert-worldpoint"* and write a_{wp}), and a list of worlds (eventually empty) of which the fact has been retracted (we will call *retraction-worldpoints* and write r_{wpi}).

The label of a fact $\mathfrak{I}$ will be thus:

$$label(\mathfrak{I}) = (a_{wp} \ (r_{wp1} \ ... \ r_{wpn})).$$

Problem: If we have an objetc A with **label(A)** = (V_1 (V_2)) and we want reinsert A in V_3 (where $V_1 < V_2 < V_3$) as in fig. 9.2:

---

[9] See definition 2.2. We remember that a proposition is *possibly true* in w_i IFF $\exists w_j \in PW$ where the proposition is true and such that (w_i **R** w_j).

[10] The label denote the set of worlds where the object is true.

Fig. 9.2.

Which will be the label of *A*? We have three possibilities:

1. Add a *special attribute* to each object to represent the worlds where this object is true;
2. *duplicate the objects*;
3. *duplicate the labels* (to have one label for each minimal world).

The first solution seems to be the most "natural" but in reality it is equivalent to the second one, which is the solution chosen by Art[11]. But the duplication of objects penalizes the pattern-matching (which is the most expensive operation of the inference mechanism) and *we have adopted the third solution*.

The label of *A* will be thus:

$$\text{label}(A) = ((V_1 \ (V_2)) \ (V_3 \ ())).$$

3 Connections with the Modal Logic

The link between $\mathcal{PW}$-*XRete* and the modal logic is made through the possible worlds (PW) semantics, and this connection is explored now to give a $\mathcal{PW}$-*XRete* a formal framework: the $\mathcal{L}_{PW}$ logic.

3.1 $\mathcal{L}_{PW}$ logic

Language The language of $\mathcal{L}_{PW}$ has as its vocabulary a denumerable set S of atomic sentences: the truth-functional connectives ¬, ∨, ∧, → and the operators □ and ◇[12]. The *well-formed formulas* (wff) are built as usually[13].

---

[11] In fact, the objects of the same class with different attributes are considered as *two different objects*.

[12] Since □ and ◇ are connected by the relations $\Box p = \neg\Diamond\neg p$ and $\Diamond p = \neg\Box\neg p$, it is enough to take one of them as primitive.

[13] We should note that all wff of the propositional calculus is a wff in $\mathcal{L}_{PW}$.

Model theory

Definition 3.1 *A $\mathcal{L}_{PW}$* **model** *$\mathcal{M}$ is a triplet $<PW, \preceq, \pi >$ where:*

- *PW is a non-empty set of possible worlds;*
- *$\preceq$ is a binary accessibility relation over PW;*
- *$\pi : PW \times VP \rightarrow \{true, false\}$ is function of assignment of truth values to the couple (w,p), where $w \in PW$ and $p \in VP$ (the set of propositional variables).*

Definition 3.2 *$\mathcal{M} \models_w \mathfrak{S}$ means that the world w of $\mathcal{M}$* **satisfies** *the formula $\mathfrak{S}$.*

Definition 3.3 *A formula $\mathfrak{S}$ is* **satisfiable** *IFF there is a model $\mathcal{M} = <PW, \preceq, \pi >$ and a world $w \in PW$, such that $\mathcal{M} \models_w \mathfrak{S}$.*

Definition 3.4 *A $\mathfrak{S}$ formula is* **valid** *(written $\models \mathfrak{S}$) if it is true il all models, i.e., $\forall \mathcal{M}$ and $\forall w \in PW$, $\mathcal{M} \models_w \mathfrak{S}$.*

Definition 3.5 *Let a wff A. The set $\{A\}^{pw}$ represents the set of worlds where A is satisfied:*

$$\{A\}^{pw} = \{w \mid \mathcal{M} \models_w A\}.$$

The truth value of a formula A in a world w (in a model $\mathcal{M}$) is defined as follows:

Definition 3.6 *Let a $\mathcal{L}_{PW}$-model $\mathcal{M} = <PW, \preceq, \pi >$ and let two formulas A and B in $\mathcal{L}_{PW}$:*

- $\{\bot\}^{pw} = \emptyset$;
- $\{\top\}^{pw} = PW$;
- $\{p\}^{pw} = \{w \mid \pi(w, p) = \top\}$, where $p \in VP$;
- $\{\neg A\}^{pw} = PW \setminus \{A\}^{pw} = Comp\{A\}^{pw}$ [14];
- $\{A \wedge B\}^{pw} = \{A\}^{pw} \cap \{B\}^{pw} \neq \emptyset$;
- $\{A \vee B\}^{pw} = \{A\}^{pw} \cup \{B\}^{pw}$;
- $\{A \rightarrow B\}^{pw} = \{A\}^{pw} \subseteq \{B\}^{pw}$;
- $\{\Box A\}^{pw} = \{w \in PW \mid \forall w_j, w \preceq w_j, w_j \in \{A\}^{pw}\}$

Finally, as a direct consequence of the preceding definitions:

$$A \models_{\mathcal{M}} B \quad IFF \quad \{A\}^{pw} \subseteq \{B\}^{pw}. \tag{9.1}$$

---

[14] $Comp\{A\}^{pw}$ is the complement of $\{A\}^{pw}$ and corresponds to the *negation by default*.

Proof theory We define $\vdash$ as a relation containing $p \vdash p$ ($\forall p \in VP$), and closed by:

Connectives	Left	Right
$\wedge$	$\dfrac{A\vdash}{A \wedge B\vdash}$	$\dfrac{\vdash A \quad \vdash B}{\vdash A \wedge B}$
$\vee$	$\dfrac{A\vdash \quad B\vdash}{A \vee B\vdash}$	$\dfrac{\vdash A}{\vdash A \vee B}$
$\rightarrow$	$\dfrac{\vdash A \quad B\vdash}{A \rightarrow B\vdash}$	$\dfrac{A\vdash B}{\vdash A \rightarrow B}$
$\neg$	$\dfrac{\vdash A}{\neg A\vdash}$	$\dfrac{A\vdash}{\vdash \neg A}$
$\square$	$\dfrac{A\vdash}{\square A\vdash}$	$\dfrac{\vdash \square A}{\vdash \square\square A}$

with:

- $A \vdash$ representing $A \vdash \bot$
- $\vdash A$ representing $\top \vdash A$[15].

Theorem 3.1 $\mathcal{L}_{PW}$ *is consistent.*

Proof: A wff A is not consistent (cf. consistent) if $\neg A$ is a theorem (cf. is not a theorem) of $\mathcal{L}_{PW}$.

$\mathcal{L}_{PW}$ *is consistent IFF for all wff A, $\{A \wedge \neg A\}^{pw} = \varnothing$.*

But:

$$\{A \wedge \neg A\}^{pw} =_{def} \{A\}^{pw} \cap \{\neg A\}^{pw} =_{def}$$
$$=_{def} \{A\}^{pw} \cap Comp\{A\}^{pw} = \varnothing.$$

Q.E.D.

Theorem 3.2 $\mathcal{L}_{PW}$ *is sound, complete and decidable*[16].

4   Industrial Application

$\mathcal{PW}$-*XRete* was developed to deal with *real life problems*. Its capabilities have been tested in military applications of Thomson-CSF such as *data fusion* or *radars identifications*. We present now a complex real problem: the Aircraft Sequencing Problem (ASP).

---

[15] The $\square_{left}$ and $\square_{right}$ rules are equivalent to the axioms **T** and **S4** of the modal logic, respectively:
- **T**: $\square A \rightarrow A$
- **S4**: $\square A \rightarrow \square\square A$.

[16] The proofs can be found in [Cav93] but we note that our accessibility relation is a *well-founded partial ordering relation* that defines a modal class equivalent to the modal system **S4** (for the proofs concerning the completeness and the soundness of **S4**, refer to [Che84] and [HC68]).

4.1 Introduction

Only in Europe, an average of 20% to 25% of the flights suffers a delay of more than fifteen minutes, causing the waste of millions of tonne of kerosene. According to IATA (International Air Traffic Association), the annual cost of these aerial bottlenecks is US$ 10 billions (US$ 5 billions in Europe and US$ 5 billions in the USA! [Jac92]) approximately.

The Air Traffic Control Problem (ATCP) can be subdivided into many minor and relatively independent problems. The most important of them is the Aircraft Sequencing Problem(ASP). In fact, a survey mentioned by [BOS86] shows that only 10% of the delays occur during the airplanes flights while 60% and 30% of the delays occur in the airplanes' landing and departures, respectively.

In the periods of rush, the control of the planes' arrivals and departures becomes an extremely complex task, yielding excessive delays that stress the passengers and air traffic controllers, waste fuel and increase the pollution[17].

4.2 The problem's resolution using $\mathcal{PW}$-$\mathcal{X}Rete$

In almost every airport in the world the strategy of sequencing adopted is the strategy of the First-Come-First-Served(FCFS), that, as is shown by [Bak74], is a strategy easy to implement but which can cause excessive delays. Our goal is, thus, to find a strategy that could provide us with an optimal sequencing, respecting all the problem's constraints.

Such a sequence must exist: theoretically it is possible to verify all the possible sequences and to choose the best one. In reality, though, this verification is impracticable. Just to give an idea of the enormous size of such a task, it is enough to consider that with just 10 airplanes we would have to effectuate **3.628.000** comparisons, whereas with 15 airplanes they grow to **1.307.674.368.000**!!

Given a set of N planes and their expected time arrival (ETA), we want to minimise the global planes delay respecting the following constraints:

1. **C1** - The Scheduled Time Arrival (STA) must be included between the earliest time arrival and the latest time arrival.
   $STA_i \varepsilon \{ETA_{i-min}, ETA_{i-max}\}$
2. **C2** - The security distance between two aircrafts (given by the matrix of security distance) depends to the order of the aircrafts in the sequence.
   $\forall i, j \varepsilon \{1, N\}, STA_i - STA_j \succeq C_{ij}$
3. **C3** - Overtakings are forbidden near the runway (in the so-called *Critical zone*)
   $ETA_i \varepsilon \{Critzone_{min}, Critzone_{max}\},$
4. **C4** - Overtakings are forbidden in the corridors which are in the *controlled zone* (zone controlled by a controller center).
   $\forall i, j \varepsilon \{1, N\},$
   $corridor(i) = corridor(j),$
   $IF\ i \prec j\ THEN\ rang(i) \prec rang(j)$

---

[17] A more detailed description of this problem can be found in [Cav92].

During traffic peak periods, the frequency of the landings is about one aircraft/minute, with a *critical zone* of about five minutes and a controlled zone of about thirty five minutes.

Matrix of security distance (s)

cat	B-747	B-727	DC-9
B-747	96	200	181
B-727	72	80	70
DC-9	72	100	70

The rules In the article [Cav92] we show that the ASP can be modeled as a flexible manufacturing scheduling and solve by A^* algorithm.

In the only rule permitting to solve the problem:

- The object *sch* represents the sequence of planes in a given world;
- The object *corridor* gives a list of planes in the same corridor;
- The object *plane*appears twice in the rule. When it is bound to the variable *?lastp* it represents the last plane scheduled in the sequence, while when it is bound to *?plane* it represents the next plane to enter in the schedule;
- The objects *table-of-security-distances* and *number-of-planes* represents, respectively, the matrix of security distances between the planes and the total number of planes that should be scheduled.

```
♣
rule scheduling priority (A* (+ ?ct ?dist) ?sch)
    (table-of-security-distances pl1=?cat tdist=?dist)
    {?corr:(corridor id= ?id corlist=?clist)
    ?sched:(seq nb-of-planes=?np ctime=?ct rang=?lrang)
    ?lastp:(plane id=?pid0 & =(car ?lrang) cat=?cat0 eta=?eta0 tot=?tot0 =?tard0 real=?real0 cor=?c0)
    ?plane:(plane id=?pid & =(car ?clist) cat=?cat eta=?eta tot=?tot tard=?tard real=?real cor=?coul
            && '(nomember ?pid ?lrang)')} in ?w0
    (number-of-planes num=?nplanes & > ?np)
    ->
    (makesubworldof ?w1 ?w0)
    {retract (?lastp)} in ?w1
    {retract (?plane)} in ?w1
    {assert (plane id=?pid cat=?cat eta=?eta real=(+ ?ct ?dist) tot=?tot
            tard= ?tard cor= ?coul)} in ?w1
    {retract (?corr)} in ?w1
    {assert (corridor id= ?id corlist= (cdr ?clist))} in ?w1
    {retract (?sched)} in ?w1
    {assert (seq nb-of-planes=(+ 1 ?np) ctime=(+ ?ct ?dist) lcreal=(newlc ?ct ?dist ?lcr)
            rang=(newrang ?pid ?lrang))} in ?w1
endrule;
♣
```

Example Let's take as an example the ASP with four planes, three categories and two corridors[18]:

---

[18] *ID* represents the identification of the plane; *ETA* denote the expected time arrival; *CAT* represent the category of the planes; *COR* represent the aerial corridor.

id	eta	cat	cor	fc_p	fc_t	pw_p	pw_t
01	006	1	1	01	000	01	000
02	012	2	1	02	160	03	140
03	038	3	2	03	260	02	210
04	051	2	1	04	330	04	320

In comparison to the FCFS solution (column fc_t), the optimal sequence found by the algorithm (column pw_t) permits a reduction of about 10% of the aircraft's delays.

An exhaustive search of the optimal solution would have made **24** comparisons, whereas the algorithm implemented in PW-XRete has generated only **8** worlds to find the optimal solution.

In the beginning the sequencing plan is empty (sch()) and we have 4 planes to be chosen. The airplane chosen was the plane *1*. The dotted lines in the picture point out the worlds that were *not* created because they were not relevant to the calculus[19] Of this initial world, two worlds are possible: the first one with the airplane *3* and scheduling *sch(1)* with priority 33; and the second one with the airplane *2 sch(1)*, with priority 31. The world that will be developed is the one with the greatest priority, i.e., the first one. The numbered circles indicate the order in which the worlds were developed.

When a rule is fired a new world is created (action *makesubworldof*). The old scheduling is retracted from and the new scheduling (the old one plus the newly chosen plane) is inserted in this new world.

Practical results Tests have been realised with real data in a SPARC-2 work-station. In practical terms, we have an average of 20 planes to be scheduled (in the rush periods this number can reach 30 or 40 planes).

The following table shows the results obtained[20]:

planes	time(s)	worlds
10	3.6	27
10	10.9	34
10	17.2	52
15	5.8	36
15	252.2	217
20	5.6	22
20	7.1	29
40	8.9	33

The differences in the CPU time (and in the number of developed worlds) for two problems with the same number of planes shows that the quantity of planes *is not* a very relevant factor. The complexity of the problem is mainly a result of the number of

---

[19] This capacity to limitate and to control the generation of new worlds will be essential in real life examples, where the number of planes is bigger.

[20] The first columns indicates the total number of planes to be scheduled; the second one shows the CPU time (in seconds) and the third one the total number of worlds developed. We remark that Fages & Jourdan have used constraint logic programming to solve the same problems with very good results [Jou92].

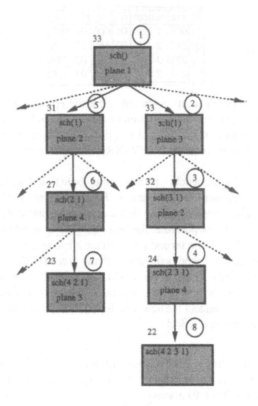

Fig. 9.3. The worlds' tree for the ASP

categories and corridors: the bigger they are the more complex is the resolution of the problem.

5 Conclusion

We have presented a system based on Kripke's possible worlds semantics for modal logic. However, for our purposes Kripke's framework is not sufficient because it is based on *monotonic* reasoning and we want to use a *nonmonotonic* one.

To take into account nonmonotonic inferences we introduced, thus, the *labeling of the worlds*. We showed that this implementation technique doesn't penalize the pattern-matching and makes easy the operations of assert/retract of an object.

Another feature of *PW-XRete* is the explicit and controlled way by which the worlds are created[21].

---

[21] The creation of a world is determined by the user and implemented by the action makeworld.makesubworldog.

The viewpoints idea from Art is similar to our labeling of the worlds. Nevertheless, their procedures duplicates the objects and penalize the pattern-matching.

Concerning the hypothetical reasoning, the ATMS are considered an obligatory reference, but they are not well suited for nonmonotonic reasoning. Moreover, their uncontrolled way of creating new contexts (contrarily to *PW-XRete* finish by restricting their application in real problems.

Finally, the examples presented showed that *PW-XRete* is a powerful metalanguage for hypothetical and revocable reasoning and that it can be used to solve efficiently real life problems.

Acknowledgements I would like to thank Christinne Froidevaux (LRI-Université de Paris XI- Orsay, France) and all the group of LACS (Thomson-CSF/LCR), headed by François Fages for their helpful comments and fruitful discussions about this work.

References

[Art86] Art. The automated reasoning tool reference manual. Technical report, Inference Corporation, 1986.

[Bak74] K.R. Baker. *Introduction to Sequencing and Scheduling*. Wiley, 1974.

[Bel90] J. Bell. The logic of nonmonotonicity. *Artificial Intelligence*, 1889/90.

[BOS86] L. Bianco, A. Odoni, and G. Szego. *A Combinatorial Optimization Approach to Aircraft Sequencing Problem*. Springer-Verlag, 1986.

[Cav92] M. Cavalcanti. Possible worlds approach to aircraft sequencing problems. *Proceedings du 1er. Workshop en Systèmes Autonomes en Robotique et CIM (Computer Integrated Manufacturing)*, 1992.

[Cav93] M. Cavalcanti. *Les Mondes Possibles dans les systèmes de production*. PhD thesis, Université Paris XI - Orsay, 1993.

[Che84] B.F. Chellas. *Modal Logic, an Introduction*. Cambridge University Press, 1984.

[dC85] L. Farinas del Cerro. *Molog: a System that extends Prolog with Modal Logic*. UPS, 1985.

[DK77] R. Davis and J. King. An overview of production systems. *Machine Intelligence*, 1977.

[Fag91] F. Fages. A new fixpoint semantics for general logic programs compared with the well-founded and the stable model semantics. *New Generating Computing*, 1991.

[Fro91] C. Froidevaux. Introduction aux logiques modales. Technical report, LRI - Université de Paris XI, 1991.

[Gab85] D.M. Gabbay. Theoretical foundations for nonmonotonic reasoning in expert systems. *NATO ASI*, 1985.

[HC68] G.E. Hughes and M.J. Cresswell. *An Introduction to Modal Logic*. Methuen and Co., 1968.

[Jac92] J.F. Jacquier. Les technologies au secours du trafic aerien. In *L'usine nouvelle*. numero 2371, 1992.

[Jou92] J. Jourdan. Modelisation of the aircraft sequencing problem in constraint logic programming. Technical report, Thomson-CSF/LCR, 1992.

[Kri59a] S.A. Kripke. A completeness theorem in modal logic. *The Journal of Symbolic Logic*, 1959.

[Kri59b] S.A. Kripke. Distinguished constituents. *The Journal of Symbolic Logic*, 1959.

[Kri63] S.A. Kripke. Semantical analysis of modal logic. *Zeitschrift fur mathematische Logik und Grundlagen der Mathematik*, 1963.

[Leh84] D. Lehmann. Knowledge, common knowledge and related puzzles. In *Proceedings 3rd. Annual ACM Symposium on Principles of Distributed Computing*, 1984.

[RFC85] M.C. Rousset, B. Faller, and M.O. Cordier. Optimisation de l'opération de "pattern-matching" dans les systèmes experts. *T.S.I. - Technique et Sciences Informatiques*, 1985.

Investigations into the Application of Deontic Logic

Nienke den Haan

Abstract. This paper discusses the results of a study for representation of law. Starting point is a legal knowledge based system for the Dutch traffic law that treats permissions as specialized obligations. The results of this prototype system have been evaluated and were the subject for further study. The second half of this paper looks into computationally attractive formalisms that capture deontic modalities more naturally.

1 Introduction

Deontic logic is used to reason about ideal and actual worlds. Its most important applications within the area of computer science are legal applications. There are however more interests (cf. [Wieringa & Meyer, 1991], [Kwast, 1991]), such as authorization mechanisms for databases, or system specifications for descriptions of permitted and forbidden states. In every case where behaviour is to be prescribed, deontic logic finds its application. In this paper the *application of deontic logic in legal knowledge based systems* will be discussed. Jones and Sergot describe in [Jones & Sergot, 1992] the problems one can encounter employing the application of deontic logic. They agree that applications of deontic logic provide extensive grounds for (future) research. This research is aimed at legal reasoning about *law texts*, since the European continental law system is *statute based*, as opposed to the Anglo-Saxon systems which are highly *case–based*. The goal of law texts is to provide a framework for general accepted or agreed upon norms.

To adequately design legal knowledge based systems (LKBS) constraints on *flexibility* and *isomorphy* must be met (see also [Bench-Capon, 1989]). Since law texts are frequently updated, such systems have to be flexible. Updates in the law texts, reinterpretations of terms and adaptations of legal reasoning should be supported. Some LKBS are based on decision tree methods, so additions or alterations of legal rules always require the decision tree to be changed. On the other hand, the representation of a law must be isomorphic to its original statement.

This makes the knowledge representation and the reasoning processes more understandable for (expert) users, and hence provides better explanation facilities. Some LKBS cannot model all necessary legal structures, such as exceptions. They have to be precompiled or require specific modelling by knowledge engineers.

Furthermore, additional knowledge is needed for the application of law: extra domain descriptions have to be provided in order to capture *common sense* knowledge LKBS would otherwise lack (see for instance the representations for various modalities in [McCarty, 1989] and his description of a language for legal discourse (LLD). In [Breuker & denHaan, 1991] a formalism was proposed based on the separate representation of world and regulation knowledge. The representations of legal rules (sections of the law) are better readable since the support knowledge necessary for interpretation is stored in the separate world knowledge base. The regulation knowledge is a direct translation of a law text, and all non-law specific terms are described in the world knowledge. The proposed knowledge representation formalism contains no special deontic operators for the representation of norms. In this article a system is presented that is designed according to this approach.

The method is tractable, performs legal reasoning in a similar way as legal experts, and the representation has a one-to-one correspondence to the original text of the regulations. In the approach described here exceptions do not have to be recognized and precompiled, because they are determined dynamically. This is a great advantage, since not all exception structures are explicit. During the application of law, implicit exceptions may occur when unexpectedly rules contradict.

The following paragraphs outline preliminary representation and reasoning formalisms which have been applied in a LKBS called TRACS (see paragraph 5). New representation attempts and viewpoints about deontic logic are reported in paragraphs 6 and 7.

2   A provisional representation formalism for rules

For the representation of obligations, prohibitions and permissions, deontic logic provides the O, F and P modalities. Since reasoning with deontic modalities is complex, in this approach we have proposed a scheme that excludes deontic modalities. The prescriptive nature of the law text is modeled using a dual world approach, in which the regulation forms the *juridically ideal world*, and cases/situation descriptions are excerpts from the *actual world*. The essence of deontic logic, which is comparing ideal vs. actual, is still the core of legal reasoning performed (see also [denHaan & Breuker, 1991]). The deontic modality P is always an exception to either F or O, and is translated to O with extra conditions. An example is given in table 10.1.

The condition of a rule contains general provisions that constitute its application grounds. The conclusion describes the intended behaviour. In this scheme the representation of permissions is geared to the application of rules explained in the next paragraph.

legal rule		representation	
number	contents	conditions	conclusions
1	$\alpha \to O(G)$	α	G
2	$\beta \to F(G)$	β	$\sim G$
3	$\gamma \to P(G)$	$\gamma \wedge G$	G

Table 10.1. Preliminary representation scheme

3 Applying rules

Legal rules can only be applied when a detailed description is given of relevant facts. Such sets, called cases, are can be composed by analyzing a real world event. This phase normally precedes jurisdiction. However this research aims only at the application of law, so a set of previously defined case facts, the situation description is provided. The situation description consists of a number of instantiated terms, i.e. a set of *case facts*. Since the conditions define the application grounds, the conditions must be compared to the set of case facts. Afterwards, the conclusions are truth-validated according to the rules in table 10.1 and the given situation. This is shown in table 10.2.

Situation	Selected Rule	Evaluation
α	1	violation
α, G	1	ok
α, β	2	ok
α, β, G	2	violation
α, β, γ	3	ok
α, β, γ, G	3	ok

Table 10.2. Evaluations

So adding a permitted action to its condition set is necessary to limit its application. Otherwise, the representation of a permission would be the same as of an obligation. Now when γ is true but G is unknown or not the case, rule 3 is simply not applicable, and the intended behaviour is not imposed.

All the regulation knowledge is universally quantified since the regulations are meant to be valid in all situations for all respective traffic participants. So given the representation of a rule:

$$(\forall)x \downarrow 1..nF \downarrow 1 \circ ... \circ F \downarrow m \to G \downarrow 1 \circ ... \circ G \downarrow p$$

where all $x \downarrow i$ are bound in one of $F \downarrow 1, ..., F \downarrow m, G \downarrow 1, ..., G \downarrow p$. In this formula the $F \downarrow i$ denote references to world knowledge (relations between variables or other predicates). Each $F \downarrow i$ may have a number of arguments containing variables $x \downarrow j$. The $x \downarrow j$ are typed variables, e.g. V:vehicle. The operators ($\circ$) may be either conjunctive ($\wedge$) or disjunctive ($\vee$). $F \downarrow 1 ... n$ describe the necessary conditions for a legal rule and $G \downarrow 1...p$ contain the juridical statements about the agent(s) in question. For example:

$\forall(V : vehicle)$

$given(T : tram) \wedge overtake \downarrow right(V : vehicle, T) \rightarrow$
  $overtake \downarrow right(V, T)$

i.e. all vehicles are allowed to overtake trams on the right. The predicate 'given' acts as an existential quantifier.

In this approach the sections of the law text are translated into the knowledge representation as accurately as possible. All the support knowledge needed to understand (complex) terms of the domain are defined in the world knowledge. Therefore, the rule base has a one-to-one correspondence to the original law text. Not only are all the terms in the regulation described in the world knowledge base, but also all the relations between them[1]. The most important relation is the subsumption relation, because testing a rule for applicability boils down to checking whether a fact in a case can be matched to the descriptions in a rule. The subsumption relations between the traffic participants from the strongly typed traffic domain is displayed in the taxonomy in figure 10.1.

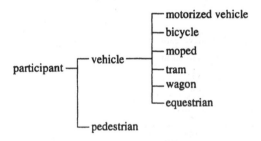

Fig. 10.1. Traffic participant taxonomy

The subsumption relations are used to determine whether e.g. a rule about a *vehicle* is applicable to the *tram* in the case. On the other hand the mutually exclusive branches can be used to determine the relation between types of negated terms[2], so a rule that states that *all vehicles but bicycles* are forbidden to use the bicycle lane is applicable to *motorized vehicles, mopeds, trams, wagons* and *equestrians*. In the traffic domain, the *predicate symbols* may also be typed as they represent traffic actions, see figure 10.2.

Types of *predicates* are necessary to determine whether the descriptions of the *traffic*

---

[1] The separated knowledge representation formalism can also be used for representing different viewpoints, e.g. when experts use different interpretations. For each interpretation a separate world kb can be used.

[2] In this approach negation by failure is used, because the situations are treated as closed worlds. In the past methods have been proposed to overcome negation in the conclusion parts of rules by proposing rule *transformations* (cf. [Kowalski, 1989]) but these harm the one-on-one relation between the representation and the original law text.

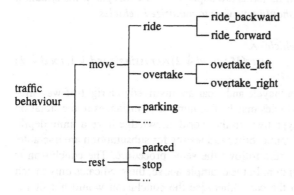

Fig. 10.2. Traffic action taxonomy

actions in rules are applicable to traffic situations, e.g. the rule 'vehicles should use (be moving in) the rightmost lane' is applicable via subsumption to a cyclist bicycling on a cyclist track. Table 10.3 gives a formalization of the determination of the applicability of rules. For this operation only the conditions of the rules need to be considered, because the conditions describe the application grounds of rules, and conclusions prescribe behaviour. Let A and B be the conditions of rules, $given(x)$ denote the presence of x in a situation description, and $\sqsubseteq$ the subtype relation in the sense of [Frisch, 1991]: a is a subtype of b, if a is in the *same type tree* as b and a is a descendant of b, e.g. *bicycle* is a subtype of *vehicle*. Negations can be proved by establishing that a and b are in the same type tree, but in sibling branches, e.g. a *pedestrian* is not a *motorized vehicle*.

$$\frac{applicable(A)\ applicable(B)}{applicable(A \wedge B)} \wedge -I$$

$$\frac{applicable(A)}{applicable(A \vee B)} \vee -I$$

$$\frac{applicable(A \wedge B)}{applicable(A)\ applicable(B)} \wedge -E$$

$$\frac{applicable(A \vee B)}{applicable(A)} \vee -E$$

$$\frac{given(x)\ (x \sqsubseteq A)}{applicable(A)}$$

Table 10.3. Application of rules

A rule is applicable when all its terms are supertypes of concepts in the situation (or equal), e.g. the rule *cyclists should give way to motorized-vehicles* :

$\forall(V : vehicle)$

    given(M : motorized $\downarrow$ vehicle) $\wedge$

    uses(V, A : road $\downarrow$ 1) $\wedge$ *uses(M : road $\downarrow$ 2)* $\wedge$ *crossing(road $\downarrow$ 1, road $\downarrow$ 2)*

       $\rightarrow$ *give $\downarrow$ way(V, M)*

applies to a situation where a bicycle and a car are involved in a right-of-way situation because the *types* of the vehicles match. The number of instances in the situation descriptions are finite and the type trees in the world knowledge have a finite depth. Therefore, the determination of applicability and testing for subsumption are tractable. Testing the conclusion part of a rule follows the same procedure. The conclusion is compared to the situation description. In the example above, the predicates only match when the bicycle gives way to the car. Otherwise the conclusion would not fit the situation description and the cyclist would violate the given rule. This way of sorted unification is based on [Frisch, 1991]. The information about types is also necessary for the determination of priorities, especially in the case of *lex specialis* which is described in the following paragraph.

4 Applying meta-rules

After the application of rules, the core of the reasoning process is *conflict resolution*. It is not very hard to determine the applicability of rules, but assigning the priority of rules is not straightforward. Rules yield conflicts when they require incompatible actions, e.g. one rule implies $O(G)$, and the other $F(G)$. Conflicts between rules arise when their conclusions are inconsistent:

$$C \downarrow 1 \rightarrow G \downarrow 1,$$
$$C \downarrow 2 \rightarrow G \downarrow 2;$$
$$\text{and:}$$
$$G \downarrow 1 \wedge G \downarrow 2 \vdash \bot$$

The clash between rule 1 and rule 2 (see table 10.1) is evident. This type of clash should never occur in a regulation. On the basis of the same set of facts, it may never be the case that conflicting conclusions can be drawn. Rule 2 and 3 do also clash, but this clash is rather relative, because a permission acts as an exception and must be solved by meta-rules. Rule 1 and rule 3 do and should not clash (cf. $Op \rightarrow Pp$).

The priority is most often determined by the meta-rule *lex specialis legi generali derogat*[3]. So when a conflict occurs, priority is given to the rule with the highest degree of specificity. This degree can be determined by looking at the conditions, i.e. the application grounds of conflicting rules. For each law domain specific meta-rules may be defined, e.g. the traffic law contains sections stating that traffic lights dominate over traffic signs, which go before traffic rules. The lex specialis meta-rule is domain independent. The meta-rules provide control over the application of primary rules. Apart

---

[3] A specific law goes before a general law.

from the common meta-rules, certain types of laws require temporal reasoning, for instance to determine whether appeals were submitted in time. For such interval handling is offered by approaches such as [Köhler & Treinen, 1994] (in this volume).

When two applicable rules result in a conflict, the *instantiated* conditions are compared. This means that only instantiated disjuncts are considered, and the instantiated condition set only contains conjunctions. The priority of $R \downarrow 2$ over $R \downarrow 1$: $R \downarrow 1 \prec R \downarrow 2$ is determined recursively over the formulas of the rules. One predicate is to be preferred over another when one of its elements (including the predicate symbol) is a supertype, and all the others are supertypes or equals of all the elements of the other predicate. If one predicate logically follows from the other, either by showing subsumption or subset relations, the order can be determined. The same procedure is followed for the complete condition sets of A and B consisting of conjuncts of predicates. Note however, that when disjunctions occur, only the instantiated disjunct, i.e. the disjunct corresponding to the situation description is used.

Consider the following example:

1: $\forall(V : vehicle)$
   $given(V \downarrow 2 : vehicle) \wedge from \downarrow right(V \downarrow 2)$
      $\rightarrow give \downarrow way(V \downarrow 1, V \downarrow 2)$.
2: $\forall(V : motorized \downarrow vehicle)$
   $given(V \downarrow 2 : vehicle) \wedge from \downarrow right(V \downarrow 1, V \downarrow 2)$
      $\rightarrow give \downarrow way(V \downarrow 2, V \downarrow 1)$.

The first states that vehicles should always give way to vehicles coming from the right. The second one explains that all vehicles should furthermore give way to motorized vehicles. The conclusions of the rules contradict, so now priorities have to be established. First of all, the predicates are equal, but here rule 2 is more specific because it mentions *motorized-vehicles*.

There are cases, where the priority cannot (easily) be established, e.g. when some terms in the condition set are less and others are more specific, or when the terms are unrelated. When rules give rise to a conflict, but a priority ordering cannot be determined, then one can either choose to select one of the rules manually, or (more important) find out *why* such an inconsistency could not be handled by the meta–rules. New jurisdiction may offer some help; sometimes a legislative error might be the cause.

5 Implementation

This research culminated in the design and implementation of a legal knowledge based system for the Dutch traffic law (see [denHaan & Breuker, 1991]). The Dutch foundation for Traffic Safety Research (SWOV) wanted a system that would automatically test versions of the new traffic law under development. Testing by hand is never complete and takes a lot of time. Another advantage of automatically applying regulations is that it is easier to acquire insight in the structure and contents of the entire set of sections. The results of the study, implementation of the prototype TRACS [4] are discussed in [denHaan, 1992]. Given a set of facts that describe a traffic situation, the system makes

---

[4] Traffic Regulation Animation and Comparison System

a graphic representation, and establishes whether any of the given traffic participants violate the traffic regulations.

The emphasis is laid on the legal reasoning rather than on the determination of the facts involved, e .g. reasoning about vague terms. The application of the rules is divided in three phases: in the first phase the rules from the regulation KB are compared to the situation using the subsumption relations in the world KB. The priorities over the resulting set of applicable rules is determined in the second phase. Interviews with the draftsman of traffic regulation made clear that the new traffic law had been remodelled and codified to match the order of the legal rules to the degree of specificity. In this system, the lex specialis has been replaced by ordering on the numbers of the rules [5]. In the third phase, only preferred rules are validated (also using the subsumption relations in the world KB) by the system in order to establish rule violations. The three phases can be seen in the legal reasoning modules in figure 10.3.

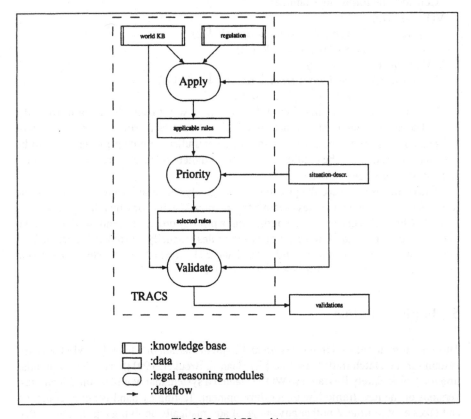

Fig. 10.3. *TRACS architecture*

---

[5] Testing the prototype system proved not only that the scope of the ordering should be limited to paragraphs only because otherwise each paragraphs would act as an exception to all preceding paragraphs, but also that the ordering within paragraphs was often wrong.

The prescribed behaviour in the conclusions of rules were easily represented in the case of obligations and prohibitions. Having a legal reasoning module that subsequently tests the applicability grounds in the conditions, and only after a positive match tests the conclusions of a rule, the deontic operators could in this case be omitted in the representation. Obligations ($O(G)$) were represented as (G), and prohibitions ($F(G)$) as ($\sim G$). The permissions gave rise to a specific problem: the permitted behaviour ($P(G)$) was not obliged, nor forbidden. Whether G is actually performed or not, permissive rules will never give rise to violations. Omitting the deontic operator for permission would give the rule the same scope and meaning as an obligatory rule. The problem was solved by examining the definition of permissions more closely. Permissions can never be violated. When the permitted behaviour is not observed in an actual situation, the norm is not violated. This notion led to incorporating the behaviour in the description of the applicability grounds, i.e. the condition part of a rule (see table 10.4). In this way, when G does not occur, the rule is not applicable, and thus not violated, and when G occurs, the conclusion part (also G) is automatically not violated as well.

legal rule	representation
$\alpha \rightarrow O(G)$	$\alpha \rightarrow G$
$\alpha \wedge \beta \rightarrow F(G)$	$\alpha \wedge \beta \rightarrow \sim G$
$\alpha \wedge \beta \wedge \gamma \rightarrow P(G)$	$\alpha \wedge \beta \wedge \gamma \wedge G \rightarrow G$

Table 10.4. Representation of rules

The solution for the representation of permissions is not very satisfactory since it is not close to the definition of law. We violate our one-to-one correspondence, and it renders tautological rules ($\gamma \wedge G \rightarrow G$). The success of the proposed representation formalism has to do with the fact that the implication ($\rightarrow$) is not used as such, because not the rules themselves are validated, but only the terms in the condition, and subsequently in the conclusion sets. So the success of the representation formalism relies on the design of the legal reasoning modules. In the next section defines features of legal reasoning and requirements needed for a more flexible representation of norms and to extend the approach to be more generally applicable.

6   General features and requirements for a generic KR formalism for law

The problems with deontic modalities were largely met by the design of a special proof theory in which the application and evaluation of rules is based on the validation of terms. The application of the prototype to traffic situations is a success, but the tautological representation of permissions is a weak spot. The general features and requirements in this section respecting general applicability of the approach have to be seen in the light of the already designed legal reasoning modules; i.e. applicability on the grounds of the condition of a rule, selection by priority, and evaluation of the conclusion part of a rule. Moreover, the initial approach did not preserve the structure of rules, a property described in [denHaan, 1994b], which afflicts the correct application of law.

6.1   Representation

The representation of a law text always calls for some application of deontic logic. The deontic logic theory however, is more extensive than is needed for the application of law, e .g. in law no iterated deontic operators are allowed (cf. standard deontic logic, see Bengt Hansson, page 124 in [Hilpinen, 1981]). However, SDL does not allow mixed formulas to represent: if *generic fact description* then *normed action*, i .e. $p \rightarrow Oq$. As example from the Dutch traffic law, section 11 .5: *Vehicles are allowed to overtake trams on the right hand side* can be represented as:

$\forall(V : vehicle)given(T : tram) \rightarrow$
$\quad permitted(overtake \downarrow right(V, T))$

In most rules, named actions only occur on the right hand side, but sometimes, for especially when also the modalities *right* and *duty* and of *power* and *liability* are embedded (cf. [Hohfeld, 1919]), a normed action can also be part of the conditions, e.g. when a citizen has acquired the right to utilization of the waterside, maintenance is obligatory. For the moment, we will concentrate on the modalities O, P and F, since they cover a large set of statutes, but the representation formalism must be prepared to contain all kinds of modalities in both conditions and conclusions.

6.2   Use of primary rules

This paragraph will discuss the role and interpretation of the implications in legal rules. In the 'classical' sense, anything can be derived from false premises: *ex falso quodlibet*. In relevant logic (or: relevance logic) the relation of consequence between conditions and conclusions has been studied (cf. [Read, 1988]). Normally:

$$\frac{\alpha, \alpha \rightarrow \beta}{\beta}$$

This is a type of natural deduction; modus ponens also holds in legal domains. However, in legal reasoning, when a rule is not applicable, its consequence is not considered because that is not relevant to the situation. Ex falso quodlibet is behaviour that at first glance must be avoided in a legal domain. Relevant logic avoids this by requiring that the conclusion must have some relevance to the condition. The relevance of a legal rule is determined by the applicability of its conditions. If the conditions do not hold, then any truth value of the conclusion may be assumed, which is ex falso quodlibet behaviour. In law this means that anything that is not specifically forbidden or obliged may be true or not in an actual situation. Therefore, standard implication *can* be used in legal reasoning. The behaviour is defined over the definition of a rule $(\alpha \wedge \beta \rightarrow G)$ and some given facts (α, β, G):

The intended behaviour in table 10.5 shows that a violation $(\perp)$ may only be established when the conditions $\alpha \wedge \beta$ hold, and the conclusion does not match the situation (G in the rule's conclusion versus given $\sim G$ in a situation).

$$\frac{\alpha \wedge \beta \rightarrow G, \alpha \wedge \beta, G}{\top}$$

$$\frac{\alpha \wedge \beta \rightarrow G, \alpha \wedge \beta, \sim G}{\bot}$$

$$\frac{\alpha \wedge \beta \rightarrow G, \sim \alpha \vee \sim \beta, G}{\top}$$

$$\frac{\alpha \wedge \beta \rightarrow G, \sim \alpha \wedge \sim \beta, \sim G}{\top}$$

Table 10.5. Legal reasoning behaviour

6.3 Exceptions

In law texts contradictions of rules often occur. The contradiction itself has no 'logical' meaning whatsoever. In fact, law texts contain numerous constructs such as:

$$
\begin{aligned}
1 &: \alpha & &\rightarrow O(G) \\
2 &: \alpha \wedge \beta & &\rightarrow F(G) \\
3 &: \alpha \wedge \beta \wedge \gamma &&\rightarrow P(G)
\end{aligned}
$$

The first rule is a general rule: normally if α then G should hold. The second rule is an exception to the first: if α and β then G is forbidden. The third rule is an exception to the second: with additional γ, G is permitted. In juridical domains the application of rules is controlled by a number of meta-rules. The meta-rules concern juridical notions about the priority of rules. It is necessary that the law text contains the third rule additional to the first, because the meta-rules would otherwise always pick the second rule as being more specific on the basis of (α, β, γ). This approach is also favoured in [Prakken, 1993]. The next section will elaborate on the meta-rules.

6.4 Juridical preference rules

The exceptions in the previous paragraph are handled by the *lex specialis* priority rule. Exception structures may be large and numerous and are often hidden, i.e. they are not clustered but scattered throughout a regulation and are only found to be conflicting when contradictions occur in the set of applicable rules. So having meta-rules to provide solutions in exception and conflict situations is to be preferred above analyzing all possible structures in advance. Then the possibility exists that not all structures, especially hidden structures, are recognized. Moreover, inserting new sections into a regulation would mean that all predefined structures would have to be reexamined. Other conflicts *lex superior legi inferiori derogat* are solved when more than one regulation is applied and so the respective statuses can differ. Where the specificity of lex specialis is determined over the subsumption relations, lex superior requires that laws from the highest legislative body get the highest priority. *Lex posterior legi anteriori derogat* can prefer a rule which is newer (posterior) over an older rule (anterior). In the Anglo-Saxon law system, where (abstractions from) cases play an important role, chronological priorities are often used. In the Dutch law system this is highly unlikely, because mostly the statutes are the only source for legal reasoning. When a new statute has become valid, the old one is explicitly invalidated. However, when instantiated rules which are derived from cases are used, time can be an discriminating factor in order to establish the priority relation. The third meta rule is the *lex superior legi inferiori derogat* which states that rules from distinctive law texts applying to the same area are distinguished with respect to the ranking of the legislative body. However, most conflicts are solved by *lex specialis*.

Sometimes the application of meta-rules does not get rid of all the conflicts in a set of applicable rules. These 'secondary' conflicts can be solved in several ways (cf. [Malt, 1992]). The conflict can be disavowed, which keeps it alive in the actual world because the conflict is not solved, or one could reinterpret on the basis of the case facts, e.g. take more arguments into account. Another solution to a secondary conflict is to declare one of the conflicting rules invalid, or not to apply a rule and so reject the intentions of the legislator. The last remedy is to reconsider the application of the meta-rules themselves.

6.5 Depth

The traffic law contains no legal rules that imply intermediate conclusions, i.e. it is never the case that the conclusion of a rule is a fact instead of a norm. Therefore, when all rules have been tested once for applicability, the set of applicable rules is complete. The case can be solved solely on the basis of the facts in the case description. When a law text contains *definitional* sections that pose intermediate facts, such facts must temporarily be added to the set of facts in the situation description.
Example:

rules:
$1 : \alpha \rightarrow O(G)$
$2 : \alpha \wedge \delta \rightarrow F(G)$
$3 : \beta \wedge \gamma \rightarrow \delta$

case:
α, β, γ, G

In this example, only the rules 1 and 3 are found applicable, so no rule is violated ($(O(G)$ matches to the case). However, rule 3 yields δ. So in a following run, rule 2 would also become applicable and would be selected by the meta-rules because it is more specific than rule 1. Eventually, a violation should be found, because $F(G)$. So after each run through the rules, the system must check whether new facts are added [6].

_____
[6] If the depth of a law text is known, this would give exact measurements on worst case time

Only when no facts are added, all applicable rules are found. The number of possible runs through the set of rules is defined as the *depth* of the law text.

Until now, examples have only mentioned rules with atomic conclusions. The next paragraph explains that problems might arise when conjunctions and disjunctions occur within the scope of deontic operators. The depth of the law requires these to be added as intermediate facts. The question arises whether the conjuncts or disjuncts can be separated so the system can use both in the line of reasoning.

7   Distribution in DL formulas

When both p and q are forbidden, does this mean that explicitly the combination of p and q is forbidden, or that both facts must be added to the set of case facts and can be treated separately? This section discusses the properties of complex deontic formulas in the light of the legal reasoning of the type discussed here. The important question is whether we can split up the arguments of a modality in order to be able to use one of its facts in subsequent reasoning. When a rule is found applicable, its conclusions are added to the set of case facts. Facts which are added to the case can now also be used in testing the applicability of other rules. Mostly, the conclusions of rules are atomic, and this yields no problem. The problems arise when the conclusion contains multiple predicates, combinations of deontic predicates and standard predicates, or complex deontic formulas. E.g. when a rule states that in case of α either p or q is forbidden, which of these must then be added to the set of case facts[7]? In table 10.6 four generic problematic types of rules are given. In rule 10.1 and 10.2 $f \downarrow 1, 2$ stand for intermediate facts which are assigned by the rules, and in the rules 10.3 and 10.4 the D denotes a deontic operator and p and q stand for normed terms.

$$\alpha \rightarrow f \downarrow 1 \wedge f \downarrow 2 \qquad (10.1)$$

$$\alpha \rightarrow f \downarrow 1 \vee f \downarrow 2 \qquad (10.2)$$

$$\alpha \rightarrow D(p \wedge q) \qquad (10.3)$$

$$\alpha \rightarrow D(p \vee q) \qquad (10.4)$$

Table 10.6. Problems with conjunction and disjunction

The status of rule 10.1 is clear. When both facts follow from the rule, they can both be added to the set of case facts. The terms in the disjunction can not both be added, because the choice between the arguments remains uncertain. So only the disjunction itself can be added.

The basic rule in standard deontic logic (SDL) is:

---

complexity of the problem solving method.

[7] Preventing disjunctions in the conclusions could be solved by splitting up into separate rules, but this violates the constraints on isomorphy.

$$O(p \wedge q) \leftrightarrow Op \wedge Oq \tag{10.5}$$

In this case we are primarily interested if the conjunction $O(p \wedge q)$ can be split up into separate facts, because we might want to *reason* with those separate facts. The question is whether the conjunction can really be split up in Op and Oq. This *depends on the interpretation* of the conjunction. If the conjunction means that p and q must hold at the same time, separation is clearly impossible. If the conjunction means that p is dependent on q or vice versa than the two facts can also not be separated. Only under the *restriction* that it is implied that both p and q are obligatory, the facts can coexist and can be treated separately.

Reversely, it is clear that $O(p \wedge q)$ can only match a set of facts that includes both Op and Oq, and that this holds for *every meaning* of $\wedge$. So here:

$$Op \wedge Oq \leftrightarrow O(p \wedge q)$$

$$\frac{Op \wedge Oq}{Op} (restricted)$$

$$\frac{Op \wedge Oq}{Oq} (restricted)$$

where the restrictions refer to the interpretation of $\wedge$. So when a conclusion consists of $O(p \wedge q)$, the conjunction $Op \wedge Oq$ can be added to the fact base, and Op and Oq as well, depending on the meaning of the conjunction. When p and q are related in such a way that they are inseparable, the definition of a new description which covers both terms at the same time might be considered, so the reasoning with $O(p \wedge q)$ is possible via $p \wedge q \equiv p'$, so $O(p')$ can be used instead.

Disjunction also causes some further examination. Does it mean that exclusively p or q must be the case, or does it mean in the classical way that only the absence of both p and q is forbidden? In SDL:

$$O(p \vee q) \leftarrow Op \vee Oq \tag{10.6}$$

So when a conclusion consists of $O(p \vee q)$ only this very fact can be added to the fact base. When at least one of Op or Oq are present in the fact base, then $O(p \vee q)$ holds. Below is showed that rewrite rules can be used to write F and P in terms of O, but carrying the distributive properties of $\wedge$ and $\vee$ over O renders some counterintuitive effects.

7.1 Rewriting F

All normative texts containing the O, F and P modalities can be rewritten to texts with only one of the three modalities. The rewrite rule from F to O is:

$$Fp \equiv O \sim p \tag{10.7}$$

When the $\vee$ distribution is carried to F via the rewrite rules, the following equation results:

$$Fp \wedge Fq \overset{(10.5,10.7)}{\leftrightarrow} F(p \vee q) \tag{10.8}$$

Now for disjunction the rewrite rules can also be applied:

$$Fp \vee Fq \overset{(10.6, 10.7)}{\rightarrow} F(p \wedge q) \tag{10.9}$$

However, intuitively one would expect that the reverse would also hold, because if both p and q are forbidden, than it can be (even with restrictions) established that Fp or Fq.

$$Fp \vee Fq \Leftrightarrow F(p \wedge q) \tag{10.10}$$

When $\alpha \rightarrow F(p \wedge q)$, then the occurrence of α forbids the combination of p and q. Singular occurrence of p or q is not forbidden!

7.2   Rewriting P

Using the rewrite rule for P:

$$Pp \equiv \sim O \sim p \tag{10.11}$$

permissions can be stated in terms of obligations. For conjunction this renders the following:

$$Pp \wedge Pq \overset{(10.11, 10.9)}{\rightarrow} P(p \wedge q) \tag{10.12}$$

Cf. the rewrite rule of prohibition above, which is what intuitively could be expected. There is no equivalence, unless the same restrictions are made as in rule (10.5). Then we have:

$$Pp \wedge Pq \Leftrightarrow P(P \wedge Pq) \tag{10.13}$$

For disjunction rewriting results in:

$$Pp \vee Pq \overset{(10.11, 10.8)}{\Leftrightarrow} P(p \vee q) \tag{10.14}$$

In the literature ([Hilpinen, 1981]) two types of P-distribution are mentioned:

$$P(\alpha \vee \beta) \Leftrightarrow P\alpha \vee P\beta \tag{10.15}$$

which is the same as rule 10.14 and obeys the rule $(P \equiv \sim O \sim)$ and the standard distributive properties. The following rule is called the notion of *strong* permission and reflects the freedom of choice principle:

$$P(\alpha \vee \beta) \Leftrightarrow P\alpha \wedge P\beta \tag{10.16}$$

The intuitive version of $\wedge$-distribution over P (rule 10.13) and the strong notion of permission (rule 10.16) both $P(p \wedge q)$ and $P(p \vee q)$ equal $Pp \wedge Pq$.

Rule	Reduction
$O(p \wedge q) \Leftrightarrow Op \wedge Oq$	Op, Oq
$O(p \vee q) \leftarrow Op \vee Oq$	$O(p \vee q)$
$F(p \wedge q) \Leftrightarrow Fp \vee Fq$	$Fp \vee Fq$
$F(p \vee q) \Leftrightarrow Fp \wedge Fq$	Fp, Fq
$P(p \wedge q) \leftrightarrow Pp \wedge Pq$	Pp, Pq
$P(p \vee q) \Leftrightarrow Pp \wedge Pq$	Pp, Pq
$p \wedge q$	p, q
$p \vee q$	$(p \vee q)$

Table 10.7. Distribution of $\wedge$ and $\vee$ over O, F and P

7.3 Distributive properties and deep law texts

The conclusion of this research is that for each deontic modality the laws for distribution of conjunctions and disjunctions are different, because if a modality is rewritten to another via the standard rewrite rules, the resulting rules for distribution are counterintuitive.

The new reduction rules (summarized in table 10.7) can be used in case of a deep law text, and enable the assertion of intermediate facts. This leaves the question of the representation of the three deontic modalities. In the next section the reader is introduced with consecutive attempts to fulfill the requirements to the knowledge representation formalism with respect to its strength and flexibility as defined in section 6.

8 Representing the deontic modalities

In this section several proposals are made for the representation of deontic modalities. To ensure that each will be tested in the same way, the following running example will be used:

$$
\begin{array}{lll}
1 : \alpha & \rightarrow O(G) \\
2 : \alpha \wedge \beta & \rightarrow F(G) \\
3 : \alpha \wedge \beta \wedge \gamma & \rightarrow P(G)
\end{array}
$$

The following situations should be evaluated as follows: When searching for *applicable* rules, the left hand sides of all rules are compared with the situation. In some situations, a set of rules is found of which certain rules clash (because $O(G)$ clashes with $F(G)$, and $F(G)$ with $P(G)$). To determine the *priority*, the instantiated rules and metarules are used. For the *evaluation* the right hand side of the selected rule is compared with the situation.

Where $\sim G$ and no G at all have the same interpretation since we use the closed world assumption[8]. The meaning of a permission is clear: when $P(G)$ it does not matter

---

[8] Bengt Hansson describes in [Hilpinen, 1981] that if f is any formula of BL, and $T(f)$ the set of possible worlds in which f is given the value truth. Let t be the set of all possible worlds. $\sim f$ is given the value true in exactly those possible worlds where f is not true, i.e. $T(\sim f) = t \setminus T(f)$.

Situation	Applicable	Priority	Evaluation
α	1	1	not ok
α, G	1	1	ok
α, β	1, 2	2	ok
α, β, G	1, 2	2	not ok
α, β, γ	1, 2	2	ok
$\alpha, \beta, \gamma, \sim G$	1, 2	2	ok
α, β, γ, G	1, 2, 3	3	ok

Table 10.8. Evaluations

if G is true, false or unknown. The next paragraphs elaborate on several attempts for the representation formalism.

8.1 The search for a representation formalism

In the prototype problems arose with the representation of the indifference of the permission. The permission behaved as if it did not matter whether G was actually the case or not, the only effect to be established seemed to be that no norm should be applied when the permission was valid. Suppose the following representation of the rules 1-3:

$$1 : \alpha \quad\quad \rightarrow G$$
$$2 : \alpha \wedge \beta \quad\quad \rightarrow \sim G$$
$$3 : \alpha \wedge \beta \wedge \gamma$$

Table 10.9. First attempt

In practice this means that when α, β, γ occur, rule 3 has still the highest priority. But *evaluation* of the rule's rhs is not necessary! There is no rhs, so indeed, no norm would be applied. The problem with this however, is that when (α, β, γ) are true now everything is allowed. This solution is adequate to solve local exception structures, e.g. table 10.9, but when a rule is a hidden exception to a section elsewhere in the statute, this exception cannot be detected. Whenever rule 3 is applicable it can serve as an exception to *any* other applicable rule. This knowledge representation formalism is clearly too weak. The scope of the permission is lost; the permission may also refer to other sections of the regulation. The scope of a permission can only be maintained when G is conserved in the conclusion of the formula. The remedy is sought by rewriting the permission to a negated rule, since $P(G) \equiv \sim F(G)$, where $F(G)$ was represented as $\sim G$. Of course the negations could not be eliminated because that would lead to an obligation structure, so the rules were translated (see table 10.10).

The advantage of this representation is that the meaning and context of the rule is still visible. In this scheme it is easy to see that the obligatory G clashes with the

$$1 : \alpha \qquad\qquad \to G$$
$$2 : \alpha \wedge \beta \qquad\quad \to \sim G$$
$$3 :\sim (\alpha \wedge \beta \wedge \gamma \to \sim G)$$

Table 10.10. Second attempt

forbidden $\sim G$. However, to establish the clash between the second and the third rule, the entire rule proposition must be compared. The second attempt will not work with respect to the consecutive validation of conditions and conclusions, because the representation of permissions can only be looked upon as one entity. The third rule in table 10.10 can be rewritten to $\alpha \wedge \beta \wedge \gamma \wedge G$ which is in essence what was used in the prototype (see table 10.4). To overcome problems with scoping and with the reasoning method that subsequently checks conditions and conclusions, G was also kept as a conclusion in the rule: $\alpha \wedge \beta \wedge \gamma \wedge G \to G$. The external negation in the third rule can be moved to the conclusion if for example two types of negations were used. Reasoning with two types of negation is however the same as reasoning with modalities. Since the representation in table 10.10 does not fit into the reasoning mechanism involved, and the former looses its scope because the norm has fallen away, the following scheme can be set up.

8.2 Using the possible world semantics

Since the essence of deontic logic is comparing an ideal world to an actual world, next a possible world representation (see for instance [Cavalcanti, 1994] in this volume) is attempted. The modalities are translated as follows:

$O(G)$: G in all worlds

$F(G)$: $\sim G$ in all worlds

$P(G)$: G in some worlds (or $\sim G$ in others)

When $G \downarrow [W \downarrow i]$ means that G is true in world $W \downarrow i$ and i ranges over all possible situations, for the running example we get:

$$1 : \alpha \qquad\quad \to \forall i : G \downarrow [W \downarrow i]$$
$$2 : \alpha \wedge \beta \qquad \to \forall i :\sim (G \downarrow [W \downarrow i])$$
$$3 : \alpha \wedge \beta \wedge \gamma \to \exists i : G \downarrow [W \downarrow i]$$

Table 10.11. Final version

The translation of $P(G)$ into disjunctive propositions to model the freedom of choice principle (i.e. $\exists i, j : G \downarrow [W \downarrow i] \vee \sim G \downarrow [W \downarrow j]$) would be weaker than the definition of contingency: G is contingent if the possible world model contains a world where G is true <u>and</u> another where G is false. Here we want to express there just might be a world where G, but that should not necessarily be our world. Rule 1 clashes with rule 2, because it can never be the case that:

$$\forall i : G \downarrow [W \downarrow i] \wedge \forall i : \sim G \downarrow [W \downarrow i]$$

Also, rule 3 clashes with rule 2, because the second rule requires all worlds to contain $\sim G$, and the third rule states that there may be a world where G. The first and the third rule do not necessarily clash.

8.3 Discussion

In this approach, the deontic aspects of law texts are captured in the comparison of the ideal vs. the actual world. With respect to deep law texts and the findings of paragraph 7 and table 10.7 the distributive properties of $\wedge$ and $\vee$ over the deontic modalities, intermediate conclusions can be stored as shown in table 10.12.

Rule	Reduction	Added facts
$O(p \wedge q) \Leftrightarrow Op \wedge Oq$	Op, Oq	p, q
$O(p \vee q) \leftarrow Op \vee Oq$	$O(p \vee q)$	$(p \vee q)$
$F(p \wedge q) \Leftrightarrow Fp \vee Fq$	$Fp \vee Fq$	$\sim (p \vee q)$
$F(p \vee q) \Leftrightarrow Fp \wedge Fq$	Fp, Fq	$\sim p, \sim q$
$P(p \wedge q) \Leftrightarrow Pp \wedge Pq$	Pp, Pq	p, q
$P(p \vee q) \Leftrightarrow Pp \wedge Pq$	Pp, Pq	p, q
$p \wedge q$		p, q
$p \vee q$		$(p \vee q)$

Table 10.12. Disjunctions and conjunctions in conclusions

Since only the conclusions of applicable, instantiated rules can be stored, the conclusions represented in the possible worlds semantics have to be instantiated as well. Obligations and prohibitions can be reduced to G and $\sim G$ respectively, but the instantiation of a permission depends on the presence of G in the situation description. In the alternative approaches in tables 10.9 and 10.10 the formulas for permissions require special reasoning. It seems that the possible worlds approach is the best option. Now the behaviour of the possible world approach with respect to the desired behaviour in table 10.8 has to be examined. Suppose we have the set of rules mentioned before table 10.8, and we apply them to the six given situations using the possible world approach. If we have the situation α, β then the set of applicable rules $(1,2)$ contains a clash because:

$$\forall i : (G \downarrow [W \downarrow i] \wedge \sim G \downarrow [W \downarrow i])$$

In general, the conflict between $O(G)$ and $F(G)$ is evident because the former requires G in *all* worlds, and the latter $\sim G$ in *all* worlds. If the set is α, β, γ, then rules 1, 2 and 3 are applicable. The conflict between rule 2 and rule 3 can also be shown:

$$\forall i :\sim (G \downarrow [W \downarrow i]) \equiv \sim \exists i : G \downarrow [W \downarrow i] \not\equiv \exists i : G \downarrow [W \downarrow i]$$

Now the meta-rules have to select rule 3 as being most specific since $(\alpha \wedge \beta) \rightarrow (\alpha \wedge \beta \wedge \gamma)$. That $O(G)$ and $P(G)$ do not clash, and also that $O(G) \rightarrow P(G)$ is visible since $\forall i : G \downarrow [W \downarrow i] \rightarrow \exists i : G \downarrow [W \downarrow i]$.

9   Conclusions

The evaluation of the method proposed in [Breuker & denHaan, 1991] has been carried out by the implementation of a legal knowledge based system. Although the system *performed* well, the results led to a reconsideration of the knowledge representation formalism. In the current approach deontic modalities will be represented in multiple world semantics. The advantages are manifold. First of all, the *representations of permissions* do not alter the definitions of legal rules as was the case in the prototype. This rather violated the envisioned isomorphy constraints. Secondly, the *detection of conflicts* can be described adequately by testing if forbidden, obligatory or permitted actions can take place in ideal worlds (combinations of the conflicting rules). Also, guidelines are provided to *solve intended conflicts and indicate unintended conflicts*, i.e. inconsistencies (see also [denHaan, 1994a]. The resulting formalism supports types of legal reasoning other than application, such as planning and advice, and consistency checking of laws. A legal knowledge based system which has been realized according to this approach can give *better pointed explanations* of the results and the legal reasoning involved (user friendliness), because the legal sources used by the system closely correspond to the original law texts. Moreover, *system maintenance* is quite clear, because the rules in the regulation base have a one-to-one correspondence to the sections in the original.

Currently the advantages of *precompiling* an isomorphic representation of law are investigated. Precompilation consists of translating the given meta-rules into the definitions of first order rules in the regulations. In a precompiled version, all the explicit and implicit exceptions have been filtered using the meta-rules. The description of the exceptions is encapsulated in the condition part of the rules. Precompilation yields a significant optimization of legal knowledge based systems.

In the beginning of this paper the research was claimed to be valuable only for *regulation based continental law systems*. The applicability of previous cases can be established by translating the information in a case into an instantiated rule. This would mean that also in *Anglo-Saxon systems* 'regulation' application can be performed: the structure of cases can be made clear by indicating new information or endorsements of existing rules and past cases.

Acknowledgements
Many thanks to Manfred Aben, Joost Breuker, Bob Brouwer and Frank van Harmelen for comments on earlier drafts.

References

[Bench-Capon, 1989] T.J.M. Bench-Capon. Deep models, normative reasoning and legal expert systems. In *Proceedings of the 2nd International Conference on AI and Law*, Vancouver, 1989. ACM.

[Breuker & denHaan, 1991] J.A. Breuker and N. den Haan. Separating world and regulation knowledge: where is the logic? In M. Sergot, editor, *Proceedings of the third international conference on AI and Law*, pages 41–51, New York, NJ, 1991. ACM.

[Cavalcanti, 1994] M. Cavalcanti. *PW-XRete: The Possible Worlds in Real Life*. In M. Fisher and R. Owens, editors, *Executable modal and Temporal Logics*, , in this volume. 1994.

[denHaan & Breuker, 1991] N. den Haan and J.A. Breuker. A Tractable Juridical KBS for Teaching and Applying Traffic Rules. In *Model-Based Legal Reasoning – Fourth International Conference on Legal Knowledge Based Systems, JURIX-1991*, pages 5–16, Lelystad, December 1991. Koninklijke Vermande.

[denHaan, 1992] N. den Haan. TRACS A Support Tool for Drafting and Testing Law. In *Information Technology and Law – Fifth International Conference on Legal Knowledge Based Systems, JURIX-1992*. Koninklijke Vermande, December 1992.

[denHaan, 1994a] N. den Haan. Eliminating Unintentional Legal Conflicts. In G. Bargellini and S. Binazzi, editors, *Proceedings of the Conference -Towards a Global Expert System in Law-*. Instituto per la Documentazione Giuridica del Consiglio Nazionale delle Ricerche, 1994. Florence, 1-3 December 1993, to appear in 1994.

[denHaan, 1994b] N. den Haan. On structure preserving representations of law. In T.F. Gordon, editor, *Proceedings of the German Conference on Artificial Intelligence and Law*, Vienna, 1994. Invited lecture.

[Frisch, 1991] A.M. Frisch. The substitutional framework for sorted deduction: fundamental results on hybrid reasoning. *Artificial Intelligence*, 49(1-3):161–198, May 1991.

[Hilpinen, 1981] R. Hilpinen, editor. *Deontic Logic: Introductory and Systematic Reading*, 1981. Reprinted, first published in 1971.

[Hohfeld, 1919] W.N. Hohfeld. *Fundamental legal conceptions as applied in legal reasoning*. Yale University Press, 1919. Fourth printing, 1966.

[Jones & Sergot, 1992] A. J. I. Jones and Marek Sergot. Deontic Logic in the Representation of Law: Towards a Methodology. *Artificial Intelligence and Law*, 1(1), 1992.

[Köhler & Treinen, 1994] J. Köhler and R. Treinen. Constraint Deduction in an Interval-based Temporal Logic. In M. Fisher and R. Owens, editors, *Executable modal and Temporal Logics*, Lecture Notes in Artificial Intelligence. Springer-Verlag, 1994. Presented at the Workshop Executable modal and Temporal Logics on the 13th International Joint Conference on Artificial Intelligence (IJCAI-93).

[Kowalski, 1989] R.A. Kowalski. The treatment of negation in logic programs for representing legislation. In *Proceedings of the 2nd International Conference on AI and Law*, Vancouver, 1989. ACM.

[Kwast, 1991] K. L. Kwast. A Deontic Operator for Database Integrity. In J.-J. Ch. Meyer and R.J. Wieringa, editors, *Proceedings of the First International Workshop on Deontic Logic in Computer Science*. DEON-91, Amsterdam, Holland, 1991.

[Malt, 1992] G-F. Malt. Methods for the solution of conflicts between rules in a system of positive law. In Bob Brouwer, Ton Hol, Arend Soeteman, Willem van der Velden, and Arie de Wild, editors, *Coherence and Conflict in Law*, pages 201–226, Deventer, Boston / Zwolle, 1992. Kluwer Law and Taxation Publishers / W.E.J. Tjeenk Willink. Proceedings of the Third Benelux-Scandinavian Symposium in Legal Theory, Amsterdam, January 3-5, 1991.

[McCarty, 1989] L.T. McCarty. A language for legal discourse I. Basic structures. In *Proceedings of the 2nd International Conference on AI and Law*, pages 180–189, Vancouver, 1989. ACM.

[Prakken, 1993] Henry Prakken. *Logical Tools for Modelling Legal Argument*. PhD thesis, Free University, Amsterdam, 1993.

[Read, 1988] Stephen Read. *Relevant Logic: A Philosophical Examination of Inference*. Blackwell Ltd, Oxford, 1988.

[vonWright, 1972] G.H. von Wright. *An Essay in Deontic Logic and the General Theory of Action*. North-Holland, second edition, 1972. ISBN 0-7204-2401-1.

[Wieringa & Meyer, 1991] R. J. Wieringa and J.-J. Ch. Meyer. Applications of Deontic Logic in Computer Science: A Concise Overview. In J.-J. Ch. Meyer and R.J. Wieringa, editors, *Proceedings of the First International Workshop on Deontic Logic in Computer Science*. DEON-91, Amsterdam, Holland, 1991.

Authors

Christoph Brzoska

SFB 314, University of Karlsruhe, P.O.Box 69 80, D-76128 Karlsruhe 1, Germany
email: brzoska@ira.uka.de

The author's paper describes work supported by the Deutsche Forschungsgemeinschaft as part of the SFB 314 (S2)

Marcos Cavalcanti

Programa de Engenharia de Producao (F-109)
COPPE/UFRJ, C.P. 68507, 21945-970 Rio de Janiero, Brazil
email: marcos@pep.ufrj.br

The author's paper describes work supported, in part, by the Brazilian Institut for Scientific and Technological Development- CNPq grant n. 202610/87.0

Michael Fisher

Department of Computing, Manchester Metropolitan University,
Manchester M1 5GD, United Kingdom
email: M.Fisher@doc.mmu.ac.uk

Thom Frühwirth

ECRC, Arabellastrasse 17, D-81925 Munich, Germany
email: thom@ecrc.de

The author's paper describes work supported, in part, by ESPRIT Project 5291 CHIC and ESPRIT Working Group 7035 LAC.

Nienke den Haan

Department of Computer Science and Law, University of Amsterdam
Kloveniersburgwal 72, 1012 CZ Amsterdam, The Netherlands
email: nienke@lri.jur.uva.nl

Jana Koehler

German Research Center for Artificial Intelligence (DFKI)
Stuhlsatzenhausweg 3, D-66123 Saarbrücken, Germany
email: koehler@dfki.uni-sb.de

The author's paper describes work supported by the Bundesminister für Forschung and Technologie under contract ITW 9000 8 as part of the PHI project.

Shinji Kono
Sony Computer Science Laboratory Inc.
3-14-13, Higashi-gotanda, Shinagawa-ku, Tokyo 141, Japan
email: kono@csl.sony.co.jp

Stephan Merz
Institute of Informatics, Technical University of Munich
Arcisstr. 21, 80290 Munich, Germany
email: merz@informatik.tu-muenchen.de

Part of the work described in his paper was prepared during the author's stay at the Institut de Recherche en Informatique de Toulouse, CNRS, Université Paul Sabatier, Toulouse, France.

Richard Owens
Nomura Research Institute Europe Ltd.
1 St Martin's-le-Grand
London EC1A 4NP, United Kingdom

Mark Reynolds
Department of Computing, Imperial College
London SW7 2AZ, United Kingdom.
email: mar@doc.ic.ac.uk

The author's paper describes work supported by the UK Science and Engineering Research Council under the METATEM project (GR/F/28526).

Ralf Treinen
German Research Center for Artificial Intelligence (DFKI)
Stuhlsatzenhausweg 3, D-66123 Saarbrücken, Germany
email: treinen@dfki.uni-sb.de

The author's paper describes work supported by the Bundesminister für Forschung and Technologie under contract ITW 9105 as part of the HYDRA project and by the Esprit Working Group CCL, contract EP 6028.

Springer-Verlag
and the Environment

We at Springer-Verlag firmly believe that an international science publisher has a special obligation to the environment, and our corporate policies consistently reflect this conviction.

We also expect our business partners – paper mills, printers, packaging manufacturers, etc. – to commit themselves to using environmentally friendly materials and production processes.

The paper in this book is made from low- or no-chlorine pulp and is acid free, in conformance with international standards for paper permanency.

Printing: Weihert-Druck GmbH, Darmstadt
Binding: Theo Gansert Buchbinderei GmbH, Weinheim

Lecture Notes in Artificial Intelligence (LNAI)

Lecture Notes in Computer Science